Philosophy of Technology

Philosophy of Technology

Frederick Ferré

The University of Georgia Press   *Athens & London*

Published in 1995
by the University of Georgia Press
© 1988, 1995 by Frederick Ferré

The paper in this book meets the guidelines for permanence and
durability of the Committee on Production Guidelines for Book
Longevity of the Council on Library Resources.

Printed in the United States of America

99 98 97 96 95 P 5 4 3 2 1

Library of Congress Cataloging in Publication Data

Ferré, Frederick.
    Philosophy of technology / Frederick Ferré.
       p.   cm.
    Reprint. Originally published: Englewood Cliffs, N.J. :
  Prentice Hall, 1988.
    Includes bibliographical references and index.
    ISBN 0-8203-1761-6 (pbk. : alk. paper)
    1. Technology—Philosophy. I. Title.
  T14.F47 1995
  601—dc20   95-10155

British Library Cataloging in Publication Data available

To Barbara

without whom it probably wouldn't have happened;

with whom it was a joy!

Contents

CHAPTER 3:
Technology and Practical Intelligence   30

CHAPTER 4:
Technology and Theoretical Intelligence   41

CHAPTER 5:
Technology and Modern Existence   54

CHAPTER 6:
Ethics, Assessment, and Technology   75

Preface

Having a chance to stand back a few cool years from the initial publication of a book can give its author a better perspective for reflection on the work than during the last hectic days of manuscript preparation. This is especially true when the book is intended not only for general readers but also as both a statement to colleagues and a help for students. The University of Georgia Press, by issuing this reprint edition, is giving me that chance, for which I am most grateful.

Several important things have happened in the interim, since I wrote this book's first preface in the freshly Chernobyl-irradiated countryside of northern Bavaria. First, from the point of view of general readers, the issue of technology has become an increasingly "hot" topic. In the years following the fateful spring and summer of 1986, when this book was written, the consciousness of multitudes caught up with what—prior to that worldwide Chernobyl catastrophe—had been the concerns of relatively few. I have been gratified at how many alert but nontechnically oriented people in my circle of acquaintances have found this book clarifying, and to that extent comforting, as they try to think about the imminent start of the next millennium and the enormous role of technology, from revolutions in genetically engineered foods to workplace and domestic implications of information superhighways, in giving shape to the civilization of not-so-long hence.

"Technology in general" has for many been too large a topic to think about with any clarity, leaving by default the alternatives of unexamined complacency or irrational fear. One of the aims of this small book has been to show that even large topics can be clarified responsibly without forbidding prerequisites. I hope it continues to serve this function in this new printing.

A second aim of the book, close to my heart (I now realize) but hardly publicly confessed, was to make a persuasive statement to my colleagues in the new and relatively small field of philosophy of technology. When this book was written there were no comparable introductions to the field as a whole. I wanted to show, by example, that philosophy of technology need not be relegated to a corner of the academic world, wedged somewhere between philosophy of science and social philosophy. Instead, philosophy of technology belongs (I believe) squarely in the center of the philosophical tradition. It unites, in scope and methods, epistemology, metaphysics, ethics, philosophy of religion, aesthetics, philosophy of history, and history of civilization; it offers more, much more, besides the history and philosophy of science.

It is doubtless unseemly to complain, because the book has been cordially reviewed, well received in many professional quarters, and satisfyingly adopted in colleagues' courses; but my hopes of drawing the whole profession toward the "humanizing" and "universalizing" of philosophy of technology so far seem unrealistically high. One event illustrates this best for me. At a special program on the book, sponsored jointly by the Society for Philosophy and Technology and the Society for the Study of Process Philosophies, held in conjunction with the Eastern Division of the American Philosophical Association in December 1988, the nontechnological speaker went for the big picture, while the speaker representing philosophers of technology was full of small-gauge doubts. I tried to respond to those doubts, of course, not only at that meeting but also through the following years by what I did with the annual *Research in Philosophy and Technology* series, of which I was general editor from 1986 until 1994. But professional lines are hard to bend; they are not elastic for those who have invested much in establishing them. Perhaps this new edition, giving new lease on life for my vision of philosophy of technology, will help shift the field of the future toward a more generous and inclusive approach. In the meanwhile, readers should be warned that what I say about philosophy of technology is not everywhere accomplished. It remains, nevertheless, my prescriptive hope.

The third major objective of this book was and is to provide a concise guide for students in classes of many sorts. I have used this book myself several times since its original publication in my "Technology and Values" course. I try to let it speak for itself; and it seems to do so without much further clarification. I use it early in the term so that standard, troubling confusions about the definition of technology and its complex relations to science can be gotten out of the way. It is short enough so that I can limit its discussion to no more than three weeks, but I can still expect students to glimpse the larger framework of ethics, society, religion, and metaphysics within which the rest of our course is spent digging in more local places.

A number of students who elect my course are rich in what they bring to the study of technology from many fields but are often beginners in philosophy. For them I have written chapter 1, to introduce them to the peculiar way of thinking that is philosophy. For others, who have already done a good bit of work in philosophy, it may be possible to plunge right in at chapter 2, which defines technology and clarifies such related issues as "natural" and "artificial." Still, the question "What is philosophy?" is a perennial philosophical question in its own right. I hope, therefore, that even advanced students of philosophy will take a look at the discussion in chapter 1, "What is Philosophy of Technology?" My main thesis is that philosophy of technology is *simply philosophy* focusing on a tremendously pervasive topic of great importance. But what philosophy *is* is not so simple, as all professional philosophers know.

I have also addressed students other than would-be philosophers. Having taught as visiting philosopher at two schools of engineering, I know how much excitement can be generated among faculty and students by the topics included in this book. This is worth a bit of work, even if it means entering new territories of the mind. The glossary was prepared to help keep certain specialized uses from being confusing. As for the rest, just let the book build. Each chapter is meant to grow from the earlier ones. Therefore I do not recommend skipping around for nonphilosophers, at least the first time through.

I have tried to strike a balance between being too abstract (for the engineers and practical people in the audience) and too absorbed in gadgets (for the theorists). Since the main need of the philosophy of technology is, as I see it, to be planted solidly on the relative permanences of the philosophical tradition, and since specific technologies are constantly coming and going, I have leaned toward my fellow philosophers in this case, though my enthusiasms as a part-time professional flight instructor and an amateur computer buff will no doubt leak through anyway. If the philosophical framework is solid and clear, each reader can supply favorite examples and specifics, now and in the future.

The book was written during a five-month leave of absence generously granted by the Regents of the University System of Georgia on recommendation of the Administration of the University of Georgia. The leave allowed me to participate in the regular exchange of faculty between the University of Georgia and the University of Erlangen in Germany. My German colleagues were kind and helpful in every respect, not only commenting on various chapters as they appeared but also defending my time for writing at home in a pleasant Frankonian village between Nürnberg and Bamberg. Colleagues in Germany and the United States who read and commented substantially on all or portions of the book were Prof. George Allan of Dickinson College; Prof. John Compton of Vanderbilt University; Prof. Don Ihde of the State University of New York at Stony Brook; professors H. J. Schneider and Christian Thiel, and Drs. R. Inhetveen and R. Kötter of the University of Erlangen; Dr. Clemens Burrichter of the *Institut für Gesellschaft und Wissenschaft an der Universität Erlangen-Nürnberg*; and professors Wyatt Anderson (Genetics) and Bernard Dauenhauer (Philosophy) of the University of Georgia. I would also like to thank other readers for their input: professors Paul T. Durbin, University of Delaware, Gerald F. Kreyche, DePaul University, and Michael Hodges, Vanderbilt University. In addition I had detailed help and welcome encouragement from Ms. Margaret Merrill, holder of the Certificate in Environmental Ethics from the University of Georgia, from Ms. Ann Causey, former member of the biology faculty at Auburn

University who is both a holder of the Certificate in Environmental Ethics and a candidate for the Ph.D. in philosophy at the University of Georgia, and from my keen-eyed mother, Mrs. Nels F. S. Ferré, a seasoned and constructive reader of family books in manuscript.

Above all I want to thank my German wife, Barbara, to whom the book is dedicated, who cheerfully has given aid and comfort to this project in every *gemütlich* way. Despite the fallout from Chernobyl—or perhaps in part because of it—Germany was a good place to think through the foundations of philosophy of technology. When even the daffodils in one's garden are radioactive, it makes one realize that there is no place to hide. For better or for worse we live in and through technology; it reflects our present social reality. Over the short run we may need to adjust ourselves to its requirements; but future social reality will be shaped, over the long run, in large part by what we believe and value—which makes this book, I hope, a potential instrument for thoughtful change.

Philosophy of Technology

CHAPTER 1

What Is Philosophy of Technology?

1.1. FOREGROUNDS AND BACKGROUNDS

We live our days—and nights—within a technological context, a "technosphere," that is immensely hard to avoid. Consider those occasions when we try to "get away from it all." Out in the wilderness, which is itself now in great part a social artifact dependent on conscious human efforts at preservation, we usually have with us a well-designed backpack, a canoe, or at least manufactured clothes, boots, matches, and pocket tools. Congratulating ourselves at our campsite on having finally slipped away from all traces of civilization, we glance up—to find the trails of a jet aircraft high above.

Normally, of course, we are content to live without much wonder inside the technosphere that surrounds us. It is as familiar as our own image in the bathroom mirror. We forget the values and knowledge that had to go into the enterprise of smoothing glass and silvering mirrors. Something so familiar as technology *in general* is actually difficult to notice, except as new or startling artifacts appear. As A.N. Whitehead pointed out, it takes a special effort of mind to pay attention to what is always or nearly always present: "We habitually observe by the method of difference. Sometimes we see an elephant, and sometimes we do not. The result is that an elephant, when present, is noticed."[1]

It is certainly important to notice such startling foreground events; evolutionary necessity has fortunately sharpened our sensitivity to them. It is no less worth-

[1]Alfred North Whitehead, *Process and Reality: An Essay in Cosmology* (New York: Macmillan, 1929), p. 6. Copyright © 1929 by Macmillan Publishing Company, renewed 1957 by Evelyn Whitehead. Reprinted with permission of the publisher.

while, however, to become critically aware of the pervasive background features of our experience as well. These, after all, are the features—looming behind every foreground event—that shape for better or worse our self-conceptions, our social arrangements, our basic expectations, as well as our sense or what is normal and normative.

1.2. THE PHILOSOPHICAL QUEST

The sustained effort at wondering critically about such comprehensive issues is essentially the philosphical quest.

All philosophy, as Aristotle said,[2] begins in *wonder*. This remains true, and those who find it painful or pointless to "waste time" in wondering about things tend to have little patience with philosophy. The special kind of wondering known as philosophy is not off-handed or sporadic, however. To do it well requires such sustained efforts of mind, such concentration and discipline, that many who have a taste for mere idle wondering also may not enjoy philosophy.

This does not mean that all philosophical thinking must result in "systems," though philosophical systems do often emerge from such sustained thinking; it does mean that philosophical wondering involves the development of mental skills and habits that do not always come easily. One main—and difficult—characteristic of philosophical wondering is that its approach is *critical* (not necessarily negative but careful and disciplined) in quality. This will be discussed later. The other great—and no less challenging—characteristic of the philosophical quest is that its issues, at least in their origin and final significance, are profoundly *comprehensive* in scope.

1.3. PHILOSOPHY AS COMPREHENSIVE WONDERING

What does "comprehensive" mean in this context? It does not mean that philosophers ignore minute issues. They often deal with them. It means, instead, that even the smallest questions of philosophy rise out of and connect back into questions so inclusive that they are extremely difficult to delimit.

a. Epistemology

One such question, for example, asks "What is knowledge?" The question does not inquire into what is *known* in any given field or by all human beings together. No matter how much that is, it would be finite quantity. Compiling it, if it could be compiled, would not be of special philosophical concern. What would interest the philosopher, instead, would be the criteria to be used in singling out what counts as the "known" and separating it from the "unknown."

[2]Aristotle, *Metaphysics* A 2, 982b12-13.

Again, the philosophical question does not ask, merely, what historical knowing is or mathematical knowing or biological or chemical or theological knowing. It asks what *knowing in general* is. What common features, if any, do all these sorts of knowing share that makes them kinds of knowing—if they do indeed all qualify as kinds of knowing? How does knowing relate to believing, to guessing, to feeling sure, to having evidence?

This question is comprehensive, then, in the sense that it asks what all knowing is like in principle. It is so comprehensive that any answer to it, to be knowable, must be applicable to itself as well as to anything else that counts as known. This set of questions, and more like them, points to but hardly "delimits" the philosophical field of *epistemology*, formed from the Greek word *episteme*, meaning knowledge, and the rich *logos* ending that means "concept" or "meaning" or "word(s)" or (as in this case) "study" of knowledge in general.

b. Axiology

Another characteristic philosophical question asks "What is valuable?" Again, the request is not for a list of all the valuable things on earth or in heaven. The philosopher would ask how—by virtue of what virtues—any of those items are qualified for such a list. What is it to be valuable or a value? There seem to be many different types of values: aesthetic values, moral values, economic values, religious values. Is there anything in common that makes them all valuable? They are certainly not all the same kind. A highly valued aesthetic object, like a perfect rainbow, may be of negligible economic value. What is of high economic value may be of dubious moral value. Acts held to be of intense religious value, like Abraham's willingness to slay his innocent son, Isaac, may be of negative moral value.[3] What, if anything, do all values have in common and what differentiates them?

Such issues point to the philosophical field of *axiology*, constructed from the Greek word *axios*, meaning excellence, plus the familiar *logos* ending. Within this general field of the study of value in general lie more specialized axiological subfields such as ethics and aesthetics. Once again, if the answers to the questions of axiology have any value, they must be self-referential in accounting for their own worth. This illustrates once more how hard it is to find any independent standpoint from which to deal with such comprehensive issues.

c. Metaphysics

A third vastly comprehensive question long asked by many philosophers, though of controversial standing with others (8.1.b), is "What is real?" There are many sorts of things: living things *qua* living are studied by biologists, behaving things *qua* behaving are studied by psychologists, material things *qua* physical are studied by

[3]S. Kierkegaard, *Fear and Trembling*, trans. Walter Lowrie (Garden City, N.Y.: Doubleday, 1954).

physicists, and so the list goes on. But philosophers have attempted to raise the question of what it is by virtue of which anything is real at all and how the various sorts of real things relate to each other.

To be real at all, for example, must something belong to the subject matter of physics? Or does human behavior indicate that there exists a non-physical soul or mind? Do numbers exist? Does God exist? These philosophical questions call for a theory of reality in general.

Aristotle called this type of theorizing "first philosophy,"[4] and perhaps just because his treatment of such questions was placed next after his treatment of the questions of physics, his book on first philosophy was called "Metaphysics," meaning *"meta-* (after)-the-*Physics."* For us, at any rate, the name, metaphysics, will be completely neutral to the vast range of metaphysical theories, some of which may be completely materialist views, claiming everything to be an expression of matter, while others may be completely idealist, claiming that nothing is real except mind. Nothing could be more comprehensive than the questions of metaphysics, which, like the equally comprehensive questions of epistemology and axiology, are self-referential since questioner and questions are part of reality.

The "comprehensiveness" of philosophical questions is shown further in that they not only inevitably include themselves in their answers but also impinge on each other. One unavoidable question within metaphysics is whether, and in what way, values exist. Do they have independent reality—are they "objective" as Plato and many other philosophers have held? Or is their status in reality only "in the eye of the beholder?" This is a metaphysical aspect of axiology.

An axiological aspect of metaphysics, *vice versa*, would raise the question of why God, if there is a God or not, should be considered good. Is it because God creates and determines all values, perhaps? Or is it because in God all values, independently determined, are present in highest measure? One's theory of values as well as one's metaphysical views will guide one's answer to these questions.

All one's answers will be subject to the epistemological challenge: "How do you know?" Metaphysics and axiology both have their epistemological aspect. And no less surely epistemology has its metaphysical aspects, e.g., in considering questions of truth. Is truth a relationship between thought and reality? What is the character of that relation? Much will depend on the status given to mind and ideas. And all this inevitably shows that epistemology and metaphysics have axiological aspects, which deal with the values of truth and knowledge (if they have value) and of the mind (if there is mind).

The comprehensiveness of its main questions, then, is one necessary condition of the kind of sustained effort at wondering that counts as the philosophical quest. No matter how detailed and technical the discussion among philosophers may become, it can be related back to these generating, omnirelevant issues.

[4]Aristotle, *Metaphysics* E 1, 1026a23-32.

1.4. PHILOSOPHY AS CRITICAL WONDERING

Philosophy is not only comprehensive in its questions, it is also critical in its methods. In its specific function as theorizing, philosophical thinking is distinguished from types of comprehensive reflection that may be engaged in for different purposes: for daydreaming, for planning a poem, for religious "global" model making, or the like.

This is not to deny that much philosophy may have its psychological origin in some or all these expressions of intelligence. Some philosophies, indeed, lend themselves better than others for non-philosophical applications of these sorts, which may suggest a close and sometimes reciprocal relation between theoretical and non-theoretical intellectual activities. Still, theorizing as an enterprise has basic internal requirements that philosophers must respect, in one way or another, if they are to wonder as *philosophers* at all.

a. Consistency

One of these essential requirements of theory is that ideas forming parts of the theory must not cancel each other by irreconcilable conflict. This is the "bottom line" value of the basic critical requirement of *logical consistency* in theorizing. To think anything in a sustained way requires the thinker to deal in some way with conflicting ideas, but not by merely asserting them on the same level all at once. In that way lies mere confusion and ruined theory.

It is true that some philosophers make a method of using paradox to make their point. Paradox, irony, metaphor, all call for effort from the reader and may, when well used, enhance the theoretical enterprise by forcing thought to run deeper than on the otherwise familiar conceptual surfaces.

But where paradox fails to point beyond itself and merely ends in unresolved contradiction, there is a breakdown of philosophical thought. This breakdown in theorizing may sometimes be sought for other motives, often religious ones, but a philosopher who desires to continue in the philosophical mode must treat a contradiction as a challenge to conceptual improvement. It is a sign of trouble and a goad to reformulation, not a resting place.

b. Coherence

A second internal need built into the theoretical enterprise is that the thinker can move from one thought to others in a relevantly connected way. This is the basis for the requirement of *logical coherence* in critical thinking. Such coherence presupposes the absence of unresolved contradiction, but it goes much further. It allows the hanging together of ideas beyond their simple co-existence.

This does not mean that the range of philosophical ideas need always to be "global." Solving a particular small issue within one modest corner of philosophy may be the defined goal. Whatever the range, however, conceptual coherence is a standard for success. If ideas do not fit together, there is more work to be done, more critical wondering about the basis for further connections.

c. Adequacy

Both logical consistency and logical coherence are standards to be applied to ideas that are being put together in a theory. Critical thought needs another standard, however, leaning to the empirical—or "data"—side, to remind the thinker that a theory needs to be rich enough in its concepts to be fair to the subject matter being wondered about. It is comparatively easy to achieve a tight coherence among ideas if all difficult or recalcitrant data have been left out at the start. Philosophers do not often mean to cheat, but the drive to coherence is so strong that sometimes there is a temptation to leave certain trouble-making experiences out of account.

Against this temptation, critical thinking insists on the standard of *conceptual and empirical adequacy*. Adequacy is difficult to achieve, since our concepts tend to influence what we notice; but our concepts and expectations do not completely dominate our capacities for noticing. Such noticing and including are immensely important for any theorizing, especially when the range of subject matter is as complex and comprehensive as in philosophy. Just as the drive to adequacy stirs the empirical scientist to look again, so it stimulates some philosophers to concentrate on providing detailed and penetrating descriptions of the structures in experience as *phenomenologists*, or students of the phenomena confronting the would-be theorizer.

Summary

The philosophical quest is defined by these two great conditions: it is the sustained way of wondering that is at once comprehensive and critical in the ways we have indicated. Of course it is difficult to hold all this together. It would be easier to be critical if the questions were more narrow—and there is a place for "sub-routines" within the philosophical "program" that allow minute questions to be asked and answered without dealing with everything simultaneously. But in philosophy one question necessarily leads to another and to another, and the comprehensiveness of the program determines the significance of the sub-routines, not the other way around. It would also be easier to be comprehensive if the standards of critical rigor were less strict—and there is room in philosophy for much division of labor between the speculators, the describers, the testers of concepts, and the analyzers of consistency. But in philosophy the critical reckoning must always be made; the high-flying speculators must eventually land—and finding the landing field will test the quality of the aerial navigation. Sometimes, alas, philosophers just circle until they run out of fuel. In this book we shall try to do better.

1.5. PHILOSOPHIES OF . . .

This book aims to clarify the foundations of philosophy of technology. Therefore it deals with a newcomer among the many sub-fields in philosophy that begin with the phrase "the philosophy of . . ."

It is worth noting here that the entire field of philosophy can be divided roughly along two major axes. The first axis has already been plotted by laying out the great questions that motivate fundamental philosophical wondering. These characteristically comprehensive questions give rise along this first axis to *epistemology, axiology, metaphysics,* and (dealing with proper critical methods in general) *methodology.* We have seen that each of these great fields of philosophical inquiry focuses but hardly "delimits" specific concerns that can be pursued with concentration but never in total isolation. Correspondingly, in familiar university philosophy programs we often find various central, often required, courses in theory of knowledge (epistemology), theory of values and ethics (axiology), theory of reality (metaphysics), and logic (methodology), as well as courses in the history of philosophy that deal with all these major philosophical concerns, sometimes developmentally across long stretches of time, sometimes systematically by way of major thinkers, and sometimes comparatively within certain important periods. Such courses all exist on the first, fundamental axis of the field.

The second axis may be laid out perpendicularly to the first. It is the axis on which the "philosophies of . . ." certain major areas of human concern may be found. Philosophers need not wonder about knowledge or values or reality in complete abstraction from the realities of religion, science, history, art, politics, education, or the like. On the contrary, the issues that are generated by these specific areas of human activity are often the most stimulating ones for the characteristic questions of philosophical wonder. Cutting across philosophy's first axis, the second axis creates a grid:

The philosophies of . . .

	religion	science	history	art	politics	law	language	mathematics	education	mind	etc.
epistemology	(1)*	(5)									
axiology	(2)	(6)									
metaphysics	(3)	(7)									
methodology	(4)	(8)									

*See text on the following page for discussion keyed to these numbers.

Each of the areas of major human concern, found above on the long horizontal axis, lends itself to being wondered about fruitfully in light of each of the comprehensive questions focused by the traditional areas on the vertical axis.

The phenomenon of religion, for example, raises questions of the first importance for epistemology. (1) Is there such a thing as religious knowledge? If so, is it of a unique sort (through mystical illumination, from specially inspired texts, or the like), or is it of a sort that has analogies with knowledge of other kinds (like scientific inferences to unseen entities, like our knowledge of other persons, like knowledge of characters in a novel or play)? If there is no such thing as religious knowledge, what is lacking in religious claims for truth that is present in other domains of life and thought?

Again, (2) religion raises deep questions for axiology, since the claim is often made that the greatest—overriding—values of all are found in the objects of religious devotion. How are such claims to be justified or set aside? How do religious values relate to aesthetic or ethical values?

Obviously, as well, (3) religion forces many of the most debated metaphysical questions. Does God exist? Do souls exist? Are there states of conscious persistence for human beings (or other creatures) after the deaths of their physical bodies?

Methodology (4) is squarely challenged by reflecting on religion. How should one define religion? Are normal logical and other critical methods out of place in dealing with religion? Are they for some reason question-begging? If so, are there any other methods to be used for responsible philosophical wondering about religion, or must philosophy simply resign from the task?

All these issues, as one must expect from the comprehensive character of the questions, are also interdependent and mutually illuminating in fruitful ways. These methodological questions (4) tie in closely with the epistemological (1); the epistemological situation (1) cannot finally be isolated from the alleged metaphysical character (3) of the subject matter under discussion; the metaphysical arguments (3) are always open to the epistemological and methodological (4) challenge; all the issues are colored by the intense value-claims (2) considered at the axiological level.

The same exercise could be conducted for each of the areas on the horizontal axis. A warning, however, is in order. Not all the basic areas of philosophical interest will be equally fully considered, in actual practice, by many of the typical courses in "philosophy of . . ." subjects. For example, most philosophy of science courses are so heavily weighted towards (5) epistemology and (8) methodology, that in a typical course there will be little or no time given to reflection on (6) axiological or (7) metaphysical questions raised by science. This seems to be a pity.

Science, fully considered, raises many profoundly important questions (6) of value, not only obvious ethical issues of many sorts but also subtle questions of the role of value judgments in influencing research commitments and shaping preferences for alternate theories. The rewarding interplay between axiology and the other comprehensive questions of philosophy should not be ignored, as they often are, by the prevailing orthodoxies in the philosophy of science.

Likewise the fascinating implications of science for theories of reality (7) should be explored in any complete treatment of the philosophy of science. Unfortunately, residual skittishness about metaphysical questions, left by aggressive but finally unconvincing Logical Positivist crusades of earlier decades (8.1.b), continues to pervade the atmosphere, particularly among those in the profession who are drawn to teaching in the philosophy of science. This is a contingent historical matter, however, and may well be passing with time. The philosophy of science since the 1960s has been in a period of rapid change.

Summary

"Philosophy of . . ." subjects are in principle, and should be in practice, *simply philosophy dealing with a special area of interest*. That means that the questions and methods are comprehensive and critical, as always, but are turned with special interest toward discovering how those fundamental questions and methods relate to a particular segment of human concern. The benefits are double. First, the areas studied—religion, science, art, history, education, law, mathematics, language, etc.—are more fully understood, thanks to applying to them the peculiar sorts of wondering that philosophers do. Second, philosophy as a whole is strengthened by its wrestle with the special problems raised in the particular fields considered.

1.6. WHY PHILOSOPHY OF TECHNOLOGY?

As we noticed at the start of this chapter, technology is so pervasive in modern times that it is almost inescapable. Frequently technological products are in the foreground of human attention, especially when they are new and large or noisy or frightening. More frequently, the technological context of our experienced world gives what Don Ihde calls "a technological texture to life":

For example, beginning with the first conscious event of the day, it is likely that the ringing of an alarm or the sound of a clock radio is our first awareness. This is followed by a whole series of interactions and uses, which may include turning off the electric blanket or turning up the heat and in either case throwing back the technologically produced bedclothes from the technologically produced bed, engaging the vast plumbing system, and entering a veritable technological jungle in the modern kitchen with stove, toaster, hot-water system, lighting, and so on. And even the philosopher takes this technological texture for granted in his or her daily use of telephone, Xerox machine, typewriter, automobile, *ad infinitum*.[5]

Here, in the all-surrounding technological environment, we find a domain of human concern and involvement that is certainly worthy of attracting a full measure of philosophical wonder. This calls for turning critical and comprehensive thought

---

[5]Don Ihde, *Existential Technics* (Albany, N.Y.: State University of New York Press, 1983), pp. 10-11. By permission of State University of New York Press.

in a sustained way on issues focused by technology. We shall see that it will warrant adding "technology" to the long horizontal axis of human concerns, analogous to religion, art, science, law, society, history, education, and the like, that can organize philosophical reflection in a fruitful way.

Oddly, this significant phenomenon has only begun to receive the sustained attention of philosophers. The history of the philosophy of technology is generally dated from the work of Ernst Kapp, in Germany, in 1877.[6] Except for Kapp's work, and an essay on the origins of technology by A. Espinas, in France, in 1897,[7] the entire field has been a development of the twentieth century. This has had the inevitable effect of giving the concept of technology a distinctly "modern" or "high-tech" slant, which, though important, may not be conducive to the best philosophical grip on the topic, for reasons that will appear in the following chapter (2.2b and 2.6).

1.7. TECHNOLOGY AND EPISTEMOLOGY

However we come to define it, technology certainly can be anticipated to involve questions of *knowledge* in important ways. The technology of a society reflects what at least some members of that society know how to do. Notably, the sorts of modern technology characteristically associated with the development of the philosophy of technology also embody theoretical knowledge gained from the sciences. All the questions that must be asked explicitly in the philosophy of science are therefore embedded in the philosophy of technology. Unpacking these complex epistemological relationships will be the task of Chapters 3 and 4.

As we shall find, technology's relation to modern science is strongly reciprocal. Modern science could not be what it is today without the precise instruments of observation, manipulation, and calculation that a refined modern technology provides. Thus it is as true to say that technology is a necessary condition for contemporary forms of science as it is to say that science is a necessary condition for contemporary forms of technology. Epistemologically, a key question may lie in the extent to which scientific knowledge itself is an artifact of our instruments and techniques.

Long before there were sophisticated technologies of theoretical science, however, human beings were using tools and following craft traditions that embodied high degrees of practical knowledge. What is the relationship between practical and theoretical reason, and how do they both relate to technology? Is one or the other primary? How does act relate to thought?

[6]Ernst Kapp, *Grundlinien einer Philosophie der Technik: Zur Entstehungsgeschichte der Kultur aus neuen Gesichtspunkten* (Braunschweig: G. Westermann, 1877).

[7]Alfred Espinas, *Les origines de la technologie* (Paris: F. Alcan, 1897).

1.8. TECHNOLOGY AND AXIOLOGY

However defined, technology certainly also involves questions of *values* in a vital way. The technology of a society reveals and embodies what at least some members of that society want (or want to avoid), and what they consider legitimate ends and means. Knowledge alone, unharnessed to human valuing, would not result in technology any more than valuing alone, lacking the necessary knowledge, could find effective embodiment. It would not be wrong, and it might be revealing, to say that technology is the offspring in *praxis* of the mating of knowledge with value, epistemology with axiology. If the principal source for our most reliable factual knowledge is science, and if the basic values of any society constitute its actual functioning religion,[8] then the technosphere we inhabit is nothing less than the incarnation in social life of science and religion.

Moreover, technology raises in acute form all the traditional aesthetic and ethical questions of beauty and ugliness, ends and means, good and evil, right and wrong—vastly amplified, often, because of the incomparable potency that modern technology has given to human decisions. More than simply traditional, some of the ethical questions in the philosophy of technology may reasonably be considered new in kind simply because of the qualitative changes wrought by quantitative considerations. But beyond this there is a further range of unprecedented ethical questions that require wise answers—urgently—since technology has opened or will foreseeably soon open genuinely new possibilities for actions about which earlier generations never needed to deliberate (6.3). As Aristotle pointed out, one can only deliberate about what is within one's power to do.[9] Since it is the nature of technology to increase the range of human powers, the associated range of questions for which human beings must assume responsibility varies with the available technology, and ethics for the philosophy of technology becomes an excitingly dynamic field for thought and application.

1.9. TECHNOLOGY AND METAPHYSICS

The implications of technology for *reality* itself are fascinating. Among them, it might be held, is the requirement that one's theory of reality somehow acknowledge what technology shows: that reality is amenable to manipulation in regular and dependable ways by human thought and decision. Or is the reverse true: that human ends and activities are in fact manipulated by technology (8.4) in determining ways?

[8] Frederick Ferré, *Basic Modern Philosophy of Religion* (New York: Scribner's, 1967), Chaps. 2 and 3.
[9] Aristotle, *Nicomachean Ethics*, Book III, 1112a30-31.

What is human nature? Are we in fact radically different from our machines (8.3)? Or can developments in the technology of artificial intelligence reveal significant truths about what we have traditionally spoken of as our "minds" or "souls?" What implications do such questions have for issues of personal self-identity, reconstitution, or the possibility of life after death?

What are the best models for representing reality as a whole? To what extent can technology provide a useful metaphor for understanding the nature of things entire? The metaphysics of mechanism borrowed key elements from the technology of mechanical clocks; are there post-mechanistic metaphysical visions (8.6) available that might influence newer technologies?

Finally, if metaphysical self-understanding and beliefs about the fundamental nature of reality make a difference, over the long run, to the way in which people relate to each other and to the world, what are the implications of new technological images for the future of society? Metaphysical theories may not merely represent the world, they may also play a part in changing it.

1.10. TECHNOLOGY AND METHODOLOGY

Some theorists hold that technology simply *is* methodology, or "technique."[10] Even if this identification goes too far, the rational organization of procedure, technique, is without doubt a profound ingredient in modern technology. Questions about method in general are therefore close to the heart of philosophy of technology.

The critical method of thinking, as we saw, requires at a minimum the avoidance of contradictions and confusions. It also aims at reaching as high a level of coherence as possible. This requires, at the outset, that the concepts to be used within the philosophy of technology be thoughtfully examined and clarified, then faithfully employed within any theories that are developed. Unfortunately, there is no ready-made, clear consensus in the field, even on such utterly fundamental concepts as "technology" itself. Therefore, if we are to think critically in a foundational way about technology and its comprehensive implications, it will be up to us to clarify and adopt conventions of our own. That will be the task of the next chapter.

The plan of this book can now be summarized in light of all that has been said so far. Chapter 2 will continue the present focus on *methodology* and critical clarification, both on the definition of technology itself and on the logical character of our job in doing that defining. Chapters 3 and 4 explore aspects of *epistemology* raised by technology insofar as modern technology rests on types of knowledge,

---

[10] Jacques Ellul, *The Technological Society*, trans. John Wilkinson (New York: Vintage Books, 1964, originally published as *La Technique ou l'enjeu du siècle* by Librarie Armand Colin, 1954).

both practical and theoretical. Chapter 5 presents four general visions of distinctively modern technology, as we exist within it. Chapters 6 and 7 then pursue principally issues in *axiology*, discussing the implications of technology for ethics and religion. Finally, Chapter 8 carries certain of these issues to the level of *metaphysics*, on which, with a speculative look to the future, the book concludes.

CHAPTER 2

Defining Technology

2.1. AN OPENING AMBIGUITY

The language of our domain reveals a tangle of overlapping and conflicting uses. We shall sort these out as we go along.

There is one major ambiguity, however, that needs immediate notice. On the one hand, the word "technology" is commonly defined by dictionaries[1] as referring to the *study* of practical arts or (by analogy with bio*logy* or anthropo*logy*) the *science* of the industrial arts. This use is clearly at home in the titles of our distinguished Institutes of Technology, but—whether recognized by the dictionaries or not—a different use (among the general public and scholars alike) is even more familiar and widespread. In this second use "technology" refers to the practical arts and products themselves. In particular, when the word is preceded by a definite article or is used in the plural, e.g., "*the technology* of an era" or "*technologies* in use by certain nations," the term does not point to second-order "studies" but to the implements, instruments, crafts, devices, utilities, contrivances, inventions, machines, artifices, tools, engines, utensils, and techniques that constitute the first-order subject matter of the institutes of technology—and of this book.

This opening distinction only slightly simplifies our task, since now we must get clear on the central character and the boundary conditions of the pervasive first-order thing itself. But is "it" a "thing" at all? Is technology even a collection of many "things," like tools and engines, utensils and machines, devices and artifacts? Or is "it," instead, more like science or religion or education, a combination of

---

[1]*The American College Dictionary*, ed. C.L. Barnhart (New York: Random House, 1962), p. 1243.

things with activities and beliefs and attitudes? Before we attempt to work out a definition of "technology" that will help us answer this and other questions, it will be useful to take a quick and exemplary look at some swirling philosophical waters that show the need for providing ourselves with some strong conceptual anchorage.

2.2. ISSUES AND DEBATES

a. Must technologies be made of matter? Does technology, as we asked in the preceding paragraph, reduce to "things?"

Argument Pro

There is plenty of motivation to answer this question affirmatively. Some, before the age of plastics, might have been willing to assert that, ideally, technologies should be made of metal. That would be obvious hyperbole, however, since even in the "iron age" of the industrial revolution, many technologies involved rubber or glass or wood or other materials. Now we boast teflon as well as formica, and many other non-metallic materials besides. Still, if the question is whether technologies need to be made of matter, the response from some might be, "What else might they be made of?"

This answer has obvious merits. It does seem strange to think about technology as being entirely disembodied. There is, to support this, a popular sense in which the technology of an era is simply the collective hardware that one could point to, weigh, and measure. It is that by which we expect the archaeologists of the far future to remember us, if at all.

Argument Con

On the other hand, an opposing view might point out that significant technological advances can be made without any new material product, merely by rearranging things or by doing things differently. Ancient Greek military technology was revolutionized by the invention of the phalanx, which is nothing but a new arrangement achieved by getting footsoldiers to link shields and spears and move in unison. European agricultural technology was greatly changed, entirely apart from any new material implements, by the introduction of the new method of crop rotation in the three-field system. Frederick W. Taylor's time and motion studies in the early days of mass production led to major economic improvements, but Taylor's inventions were not embodied in steel or rubber: they were simply carefully revised patterns for workers to spend their energies differently.

Argument Pro

Still, the defenders of matter might reply, the Greek phalanxes; the European croplands, and the workers in the factories, were all material realities. Insofar as these were examples of technologies, they were manifested in the material world. Not all good ideas for doing needed things, after all, need to be technologies. For example, the invention of the zero and of Arabic numerals made doing mathematics enormously easier, and it might even have been practical to be able to do mathematical calculations "in the head" without needing the ground or wax tablets or paper to write on, but it would be stretching things to call such inventions and disembodied techniques "technologies" of mind or thought.

Argument Con

Perhaps so, might come the retort, but even if material embodiment—somehow—is a necessary condition of technology, it certainly is not sufficient for defining it. A tool is never mere matter. It is matter designed by intelligence for purpose. Even "found" tools, like the unsharpened stone used by our most primitive ancestors, were matter selected by intelligence for purpose. Much less is the high-tech instrumentation of modern society merely matter. It is above all scientific understanding put to use. Matter may or may not be essential to technology; intelligence clearly is.

b. Must technology be science-based? It might appear, considering this last comment, that technology, properly so-called, is necessarily dependent on the intelligence of scientists.

Argument Pro

This would accord well with the sharp distinction that one might find in popular usage between technology, as such, and mere "crafts" from pre-scientific times. Not only in the popular mind, but also in philosophic literature, there is support for this position. As Mario Bunge writes:

In past epochs a man was regarded as practical if, in acting, he paid little or no attention to theory or if he relied on worn-out theories and common knowledge. Nowadays, a practical man is one who acts in obedience to decisions taken in the light of the best technological knowledge. . . . And such a technological knowledge, made up of theories, grounded rules, and data, is in turn an outcome of the application of the method of science to practical problems.[2]

Argument Con

On the other hand, it might be replied, there is a common usage in which it is possible to speak of the varying "technologies" of different times and cultures beyond the narrow span of modern science. In popular speech one can sensibly make comparisons between contemporary rocket technologies, for example, and rocket technologies of the ancient Chinese. If technology were essentially dependent on science and the "method of science" as we know it today, then we should have to deny to other times and civilizations—wherever modern science was unknown—the acquisition of "technology" altogether. This seems unnecessarily restrictive.

Such a tight linkage between science and technology is explicitly rejected by philosophers, too, like James K. Feibleman, who writes:

Speaking historically, the achievements of technology are those which developed without benefit of science; they arose empirically either by accident or as a matter

---

[2]Mario Bunge, "Toward a Philosophy of Technology," in *Philosophy and Technology: Readings in the Philosophical Problems of Technology*, eds. Carl Mitcham and Robert Mackey (New York: The Free Press, 1972), p. 62. From Chapter 11 in *Scientic Research II: The Search for Truth*, Vol. 3, Part II of Studies in the Foundations, Methodology, and Philosophy of Science (Berlin, Heidelberg, New York: Springer-Verlag, 1967). By permission of Springer-Verlag.

of common experience. The use of certain biochemicals in the practice of medicine antedates the development of science: notably, ephedrine, cocaine, curare, and quinine. This is true also of the prescientific forms of certain industrial processes, such as cheese-making, fermentation, and tanning.[3]

Such a position would not deny that technology is now aided and greatly accelerated by infusions from applied science, but technology, from this point of view, should be defined independently of science, with which it interacts. "Technology might now be described as a further step in applied science by means of the improvement of instruments. In this last sense, technology has always been with us; it was vastly accelerated in efficiency by having been brought under applied science as a branch."[4]

Argument Pro

Against this position, the defender of the essential science-dependence of technology might well point to the radical differences between science-driven technology and the prescientific crafts. The admitted differences in efficiency between prescientific folk arts and true scientific technologies are so great, it may be argued, that they justify our insisting on a different name for them. Sometimes matters of degree get to the point where a genuine difference in kind needs to be christened with a new name.

Argument Con

In rebuttal, it may be answered that historical continuities as well as discontinuities demand attention. Insisting on a different word may obscure the important lines of continuous development between, say, certain hand tools and machine tools that developed from them. What is needed, it may be argued, is greater appreciation of the rich debt we owe our inventive ancestors instead of less. Making, by definition, the "technological" phenomenon a purely modern, science-dependent concept is risking the severing of important connections, not only comparative connections with other civilizations but also historical connections with our own roots. The consequences of such conceptual narrowing are cultural ethnocentrism and alienation both from the rest of the world and from all other ages.

c. Should technology be credited to animals? If equating the range of technology with applied science threatens us with such impoverished appreciation for our pre-scientific ancestry, then should we turn in the opposite direction and find continuities all the way back to pre-human forms of life?

Argument Pro

This is surely a promising suggestion, some might hold, if we are to be consistent. If the beaver builds dams with skill and determination, and enjoys the benefits—in food supply, dwelling space, refuge—of the resulting pond, why should this not be termed a form of technological activity? Since the animal world is full of complex

---

[3]James K. Feibleman, "Pure Science, Applied Science, and Technology," in Carl Mitcham and Robert Mackey, *Philosophy and Technology*, p. 36. From the *Two-Story World* by James K. Feibleman (New York: Holt, Rinehart and Winston, 1966). By permission of Holt, Rinehart, and Winston.

[4]*Ibid.*

contrivances, e.g., nests and hives and disguises, etc., would it not be sheer anthro-pocentric prejudice (some call it "human chauvinism"[5]) to draw the conceptual line for technology at the human race? Lewis Mumford lends support to this line of reasoning:

In any comprehensive definition of technics, it should be plain that many insects, birds, and mammals had made far more radical innovations in the fabrication of containers than man's ancestors had achieved in the making of tools until the emergence of *Homo sapiens:* consider their intricate nests and bowers, their beaver dams, their geometric beehives, their urbanoid anthills and termitaries. In short, if technical proficiency were alone sufficient to identify man's active intelligence, he would for long have rated as a hopeless duffer alongside many other species. The consequences of this perception should be plain: namely, that there was nothing uniquely human in early technology until it was modified by linguistic symbols, social organization, and esthetic design.[6]

Argument Con

Against this position it might be retorted that taking technology so compre-hensively obscures too many significant differences in the subject matter. There is all the difference in the world, it could be argued, between an "urbanoid" anthill and the actual urban human construction. Hives and anthills are purely natural. A city, on the other hand, depends on the anticipation, and at times the deflecting, of otherwise natural happenings. To divert attention from the striking differences in kind between human technology and the complex but instinctive activities found in the world of nature is to lose sight of the main point. Technology changes and often fights nature; technology introduces artifice into the world. Defining tech-nology so that there is no difference left between the natural and the artificial, so it might be argued, defeats all clarity of thought on the subject.

d. Is technology necessarily unnatural? A positive answer to this question may seem trivially obvious to some.

Argument Pro

Here, it may be argued, is the home ground of the use of the term "artificial." If technological phenomena are not artificial, thus "unnatural" in the most literal sense of this adjective, then nothing qualifies for the term.

Argument Con

On the other hand, it might be replied, there is nothing unnatural about horses or oxen, and if at a certain stage of technical development the muscle power of such animals is exerted for the achievement of human purposes, what is unnatural about that? Or what is unnatural about water flowing down a riverbed and turning a mill wheel? Is wind made unnatural when it is caught in the sails of the windmill?

---

[5]R. and V. Routley, "Against the Inevitability of Human Chauvinism," in *Ethics and Problems of the 21st Century*, eds. Kenneth E. Goodpaster and Kenneth M. Sayre (Notre Dame, Ind.: University of Notre Dame Press, 1979), pp. 36–59.

[6]Lewis Mumford, "Technics and the Nature of Man," in Mitcham and Mackey, *Philosophy and Technology*, p. 78.

Is a sailboat unnatural when it catches the prevailing wind? Does it become suddenly unnatural when, with the help of a keel or centerboard and rudder it uses friction to beat against the wind's direction? Is fire unnatural when lit by matches in a hearth but not in a forest fire set by lightning? Is it somehow unnatural to burn a lump of coal but not a log of wood? Where does one draw the line?

Argument Pro

Perhaps these questions can be answered, in general terms, by insisting that whenever human beings intervene deliberately in the world of nature, they introduce artificiality. This does not mean that there is anything "fake" or unreal about the interventions of our species. If the forces and materials of nature are being used and redirected by human beings, this alone does not make them any less the forces and materials of nature; but there is an important sense in which nature is being interfered with, changed from what it would otherwise have been. Human actions create history where there were only natural processes before.

Argument Con

In rebuttal, it might be argued that this view exaggerates the importance of the human presence within nature. Birds "use" the principles of aerodynamics to fly, but one does not say that their flight is in any way artificial. When human beings use same principles to lift their own heavier-than-air bodies through the skies, is this not a similar expression of the natural? Birds built nests to shelter their young. When human beings put materials together for the same function, why should one say in the second case but not in the first that there has been an "intervention" into nature instead of simply another manifestation of it? Indeed, this contrary position might argue, the activities themselves of building shelters—and aircraft— are expressions of a specific nature: *human* nature, of which what we proudly call the "technological" is simply the natural outcome.

e. Is technology wholly natural? The outcome of the previous exchange might suggest (to some) a firmly positive answer to this question.

Argument Pro

After all, we may be reminded, the "natural" includes everything that is except— if such there be—the "supernatural." Since technology is firmly rooted in the laws of nature, using the raw materials of nature, and since it springs out of human nature, it must be wholly natural. Despite romantic exclamations about the "miracles" of modern technology, the technological phenomenon is squarely in the domain of this world. In fact, the achievements of technology have been able to replace interest in or hope for miracles themselves. To a great extent the popular religious faith of modern civilization has been "naturalized" by the advent and advances of technology.

Argument Con

However persuasive this argument may be, it will not satisfy those who continue to insist that important conceptual distinctions are in danger of being lost by blanketing everything with the same "natural" label. Taken too far, this line of reasoning would lose the concept of the artificial and would in the process obscure the striking differences between nature when left alone and nature when manipulated by intelligence for human ends. Perhaps human intelligence is "natural" in one sense; but in another it has brought about much that would never be found in nature without its intervention. There are now literally new elements and materials that exist in the world only because of the intervention of physical and chemical technologies. There are not only new species of domesticated plants and animals but also wholly new life forms, thanks to biological technologies. The word "artificial" as antonym to "natural" may be too clumsy. A new variety of "black" tulip, carefully developed for utmost darkness of hue, is the product of artifice (skill, intelligence, etc.) but is not "artificial" in the way that a silk tulip is artificial. Still, though living, the black tulip is not completely "natural," either. More conceptual refinements are needed. We need to be able to think clearly about the new possibilities and new threats within nature that human minds, through technology, have introduced. Calling *everything* "natural" dulls our powers to distinguish in thought, notice in experience, and judge in moral terms the many ways that the human enterprise impinges on nature.

Summary

The main purpose of this exercise in unresolved disagreement has been to illustrate the need for more conceptual refinement. So long as our key terms, like "technology," "nature," "artificial," and the like, are left without adequate analysis, debates like these will merely swirl. If we are to get much beyond this level, we shall need to resolve certain issues by making crucial decisions on the limits of our concepts. Then we can return to these issues, and others, to see whether we can resolve them to our satisfaction.

This process of decision making can be entered blindly and done badly, or it can be entered with full awareness of the nature of the process and the implications of our decisions. In the interests of doing our defining well, we shall need to take a short excursion through the logic of definitions in general.

2.3. THE VOLUNTARY LOGIC OF DEFINITION MAKING

A primary point to remember in reviewing the logic of definition making is that for philosophers it is a process not of *reporting* but of *deciding*. This may seem obvious, but overlooking it often traps people, especially those new to philosophy, who confuse the philosophical process of laying down conceptual rules with the lexicographical process of dictionary making. Dictionaries attempt to report usages, and philosophers are grateful for the work of lexicographers, since they provide much raw material that needs (as we shall see below) to be taken into account. But where dictionaries stop, philosophers begin.

Frequently, for example, dictionaries will report several uses, not all of which can be used at once without contradiction. The term "nature," in one typical dictionary, provides fourteen different uses, including both "a primitive, wild condition; an uncultivated state" and "the universe, with all its phenomena."[7] These are obviously inconsistent, since if I assert the sentence "Paris is part of nature" in the first sense, my assertion will be false, but if I assert the sentence in the second sense, the assertion will be true. When this happens, philosophers need at least to choose among these uses, or stipulate some new use, for the sake of consistency. At other times, dictionaries will need to be argued with, as we did in confronting the "opening ambiguity" of this chapter, for the sake of comprehensiveness. Hardly ever will a philosophical problem be resolved, much less solved, merely by looking up some words.

Because defining a term is making decisions about how one's words are going to be used (or, since language is social, making recommendations about how words regularly ought to be used), the standards we use to evaluate a definition will be those appropriate to decisions and recommendations, not descriptions and reports. The *description* of something is "accurate" or "inaccurate," a *report* is "true" or "false"; *decisions*, on the other hand, are neither accurate nor inaccurate, neither true nor false. If I decide that I am going to read books, as a rule, rather than to watch television, this *decision* itself is not subject to such adjectives as "true" or "false," though *reports* about it (including my own reports) may be. Decisions and recommendations are, instead, wise or foolish, useful or self-defeating, effective or silly.

Since these are the sorts of standards we need to apply, it follows that associated questions can always be raised: "wise" *under what circumstances*? "useful" *for what purposes*? "effective" *within what limits*? If I recommend, for example, that you, too, always should read books rather than watch television, you might well ask whether this is wise or silly advice, but not whether it is true or false. While you are at it, you might ask whether there are circumstances (e.g., obtaining the latest news) in which a different recommendation might be more appropriate. There are, after all, different purposes to which different recommendations are suited.

Linguistic recommendations are logically similar. Laying down the rule, for example, that under certain circumstances we should use "nature" to mean "the universe, with all its phenomena," might be good for consistency and comprehensiveness, if those are our primary purposes in a certain context, but requiring that all appearances of the term in all contexts have only this use might be foolish. We would need to find new ways to talk about "getting back to nature," or "preserving nature." *Definitions are context- and purpose-dependent and should not be allowed to tyrannize their makers.*

This observation is in keeping with some famous remarks of Ludwig Wittgenstein[8] on the functional character of linguistic usage. Wittgenstein pointed out that many words (his example was the word "game") resist clear and final definition by any single identifiable set of necessary and sufficient characteristics. Instead, a loose but

---

[7]*The American College Dictionary*, p. 810.

[8]Ludwig Wittgenstein, *Philosophical Investigations,* eds. G.E.M. Anscombe and Rush Rhees, trans. G.E.M. Anscombe (Oxford: Basil Blackwell, 1953).

discernible grouping of characteristics, no one of which is always present, might be enough to ground the concept, in the way that certain features, like eye color or nose shape or gait or body build, might provide the basis for recognizing a family resemblance.

Some philosophers, perhaps overinterpreting Wittgenstein, conclude that the job of defining terms is somehow less valuable or less needed because of these contextual realities. This might be true if the aim of definition were to achieve an "absolute" rule for the use of words. We can see, however, that such a goal is not appropriate. More modest goals, and the criteria appropriate to them, will be enough for us.

2.4. TOWARD AVOIDING CONFUSION

Confusions can creep into thought either when words are used in ambiguous and changing ways or when words are used too far removed from familiar common uses.

a. Attention to Explicitness

To counter the first of these sources of confusion, explicit statements about what a term will be taken to mean in a given context—in an argument, an article, a whole book—serve the purpose of helping both the reader and the writer of a text avoid confusion by reminding them of (at least) what range of features undergird the recognition of the "family resemblances" that hold a subject matter together. In some specialized contexts it may be possible to stipulate more exactly which of these traits are those that cannot be absent whenever the term is properly applied (the necessary conditions) and what combination of these traits will guarantee the applicability of the term (the sufficient conditions). The degree of exactness needed, of course, is a function of other purposes at hand. Then, having explicitly stated the conceptual connections of a term, either by clarifying its family resemblances or by offering a formal definition by necessary and sufficient conditions, the writer has (in a manner of speaking) given a contract to the reader that the word will faithfully be used in only this manner; and the reader, by continuing to read, has implicitly accepted the contract, if only for the duration of the present context.

b. Attention to Common Usage

If the goal of avoiding confusion is taken seriously, however, the writer needs to observe some limitations on the arbitrariness of the decisions being made about terms, even within limited contexts. Each context is located within a wider context. To offer the reader a "contract" to use a term in a way entirely out of keeping with customary usage is to court confusion, however explicitly the proposed eccentric use is spelled out. Readers are not able, even with genuine good will, to make their minds blank when they start a new article or a new book. Therefore philosophers, if they value the avoidance of confusion, will need to be attentive to the work of lexicographers and the dictionary makers who report on the popular uses of terms. Defining "nature" as "any paved area of land," in consequence, would be a silly policy, even though it might be explicitly made and consistently followed, since it would be so likely to lead to "cognitive dissonances" and avoidable confusion.

It is not always possible to follow popular usage, as we have seen, since popular usage is often quixotic, incoherent, and ambiguous. The philosopher, for the sake of critical thinking, must continue to be free to choose, decide, and propose. But each freedom bears its corresponding responsibility. The responsibility of philosophical choice is that it be subject to limits set by its own goals. One such goal, the avoidance of confusion, requires that ordinary language, though not sovereign, be respected. This means at least that all our specialized stipulations should be able to be located securely on the linguistic ground that is the common starting place and final context for all such specializations.

2.5. TOWARD AVOIDING EXCESSIVE BREADTH

Since the function of a definition is to make a concept definite, it is important that definitions should be defended against including so much that everything in general—but nothing in particular—is referred to. There can be no hard and fast rule for what counts as "excessive" breadth, since all definitions are purpose- and context-dependent, but given our philosophical purpose of critical thinking, it will be clear that a useful definition in the present context will attempt judiciously to keep some things in focus for examination and let others drop into the background.

a. Technology Is Implemented, Not "Empty Handed"

One implication of this for defining technology is that we must be ready to admit that some important and related subject matters may not be part of our concept. However important our empty hands are, and have been in the history of our race, for manipulating the environment to our ends, it would be wise to resist a definition of technology that includes empty hands as technological implements. The totally naked human body, interacting face-to-face with the environment, unmediated by any artifact, contrivance, invention, or tool, would seem to stand as a paradigm case of the non-technological.

Lewis Mumford argues to the contrary that the "mind-activated body" is itself a "primary all-purpose tool,"[9] but this suggestion, if adopted, would leave us unable to talk without contradiction about people lacking tools or about the differences made to minds as well as bodies by the introduction of technics. Broadening the definition of technology to include the empty hand also puts sufficient strain on ordinary usage as to pose a real threat of confusion.

[9]Lewis Mumford, "Technics and the Nature of Man," in Mitcham and Mackey, *Philosophy and Technology*, p. 78.

b. Technology Is Practical, Not "For Its Own Sake"

The notion of the "practical" will be elaborated in Chapter 3. For now it will be enough to rely on its common meaning, supporting such ends as survival, health, comfort, and material well-being. Departing from this meaning, however, there might be another overly-broad definition of technology that would equate it, because of its Greek root, *technē*, with *any* "art or skill." There are skills, however, including skills in using instruments, that a useful definition of our subject should decide to exclude. These are the skills that are *done for their own sakes*, as in the "fine arts" as opposed to the "applied arts." It would doubtless be logically possible, were we to opt for it, to include all the fine arts, including dance, song, painting, poetry, sculpture, etc., under "technology." This decision would, however, obscure too many important differences. There are indeed technological "interactions" with the fine arts, such as important technological development in musical instruments. But the fine arts themselves should be distinguishable in concept from the technologies they may employ if there is to be any clear sense made of "interaction" between them.

A similar consideration rules out including those skills, such as religious rituals, that are engaged in for worship instead of for any mundane practical end. The ritual of the mass has been developed to a fine pitch of technique, but it would be odd to include such technique within the concept of "technology." Only as a metaphor— and perhaps a polemical one—would one speak of a priest as engaging in "religious technology" during a ceremony of worship. The Buddhist prayer wheel, like some employments of the modern television studio, is technology put to religious use. Indeed, religious motivation may have led to the invention of the wind- or water-driven wheel, which would make wind- or water-powered mills an important technological offspring of piety, but to make these cross-connections it is useful first to maintain the separation in concept. Without initial distinctions it is much harder to achieve a clear mapping of connections.

c. Technology Is Embodied, Not "In the Head" Alone

Likewise, it would be wise to guard against the absorption of all methods and techniques, including even wholly mental ones, into the concept of technology. A natural language, for example, is vitally important for virtually everything that happens in a culture, including (among much else) its practical achievements. But because of this near universal relevance we would stretch the concept of technology to the tearing point if we were to include natural languages within it. Similarly, mathematics undergirds practical life in innumerable ways; but although the technique of doing sums in one's head may be *related* to technology it should not itself be *equated* with technology.

It is clearly a different matter when we are confronted with a desktop calculator or an abacus or, indeed, even with pencil and paper embodiments of mathematical processes. These are externalized practical means to our ends. In similar ways we may consider the printing press, which is an obvious paradigm of technology, as a means to the expression, dissemination, and preservation of natural language. So also are the hieroglyph, the recording tape, the word processor, the software program, the papyrus scroll, the library, and the microfilm reader. The crucial difference between non-technological and technological in these cases seems to lie between the activity of intelligence alone and the embodiment, somehow, of that activity.

d. Technology Is Intelligent, Not "Blind"

Given the foregoing discussion, intelligence itself seems to emerge out of all this as essential to the concept of the technological. "Skill" or "art" (*technē*) is not literally attributable to anything that occurs blindly, without some degree of calculation of means to ends. If this is so, then the concept of technology will not usefully be extended to behavior that, among humans, is merely accidental or, among other species, is entirely instinctive. Making this decision would not necessarily draw an anthropocentric line isolating human beings as the only genuine manifestors of technology. It may be that significant technological expressions of intelligence are to be found among other species. This lively question is being dealt with by investigators in ethology and animal psychology. The concept of technology, once limited to manifestations of a significant degree of intelligence, however, can be given formally clearer though still empirically arguable boundaries. It will at least exclude products of instinct, like the hives of bees, and will help us concentrate on what both *technē* and *logos* point us to.

These preceding considerations would suggest that it would be wise, in attempting to shape a usefully definite conception of technology, that the concept *not* include activities or skills that are (i) empty-handed, (ii) self-justifying, (iii) disembodied, or (iv) unintelligent. Put positively, it suggests that our definition will need to stipulate that technology involves (i) implements used as (ii) means to practical ends that are somehow (iii) manifested in the material world as (iv) expressions of intelligence. In coming this far toward giving definiteness to the range to be given our primary concept and in giving reasons for our decisions, we have accomplished a large share of the necessary task.

2.6. TOWARD AVOIDING EXCESSIVE NARROWNESS

The other side of the task is making sure that our concept is not made so narrow, by over-restrictive decisions, that our philosophical enterprise of thinking *comprehensively* as well as critically is inadvertently handicapped. Once again it must be stressed that others may come to different decisions without being "wrong"; the appropriate justifications for our decisions must rise out of our purposes.

Since our basic purpose is to think about what technology *in general* means for the human race in general, pursuing the great issues of the philosophical quest, then it would seem self-defeating to isolate contemporary science-based technology from the practical manifestations of intelligence in other times and cultures. This would be to draw the boundaries of our concept so tightly around our own present culture that it would deflect our attention from all others. There are, to be sure, highly significant novelties introduced by science into the technologies of our modern civilization. But these novelties are subject to exaggeration if we set up our concepts to suggest that the technology of our time is a wholly different phenomenon from the implemented practical expressions of intelligence of other cultures and times. Thus it might be wiser not to define technology as necessarily rooted in science. Instead, familiar expressions, such as "high technology" or "scientific technology," are already at hand to distinguish our characteristically modern forms of technology from all the rest.

2.7. OUR DEFINITION

The work of fashioning a definition is mainly done once the preliminary resolutions of inclusion and exclusion have been thoughtfully made in light of an explicit purpose. We shall respect ordinary usage, while tightening its looseness, by stipulating that "technologies" in the context of this book shall mean *practical implementations of intelligence.* The word "technology," as we are using it (2.1), is simply a general term for "technologies." "Practical" requires that they *not be wholly ends in themselves*; "implementations" entails that a technology be somehow *concretely embodied*, normally in implements or artifacts, sometimes simply in social organization; "intelligence" will receive detailed discussion in Chapters 3 and 4.

This definition will *not* absolutely restrict technologies to human culture, though they will be found mainly in human cultures if purely instinctive phenomena are to be excluded.

This definition will *not* restrict technologies to contemporary scientific expressions of intelligence, though it will suggest that we pay special attention to the scientific characteristics of the intelligence that lies behind the pride and terror of our modern civilization.

This definition *will* restrict technologies to the dimension of means, though we must later discuss the extent to which technical means have tended in our culture to become also ends in themselves.

This definition *will* restrict technologies to something embodied in artifacts, though by emphasizing their origin in intelligence, it will not reduce technology to the "stuff" or matter of the artifacts alone.

These and other consequences of our conceptual decisions will become clearer as we continue the philosophic quest. It will only be through such later discussions that we can discover whether we have done our methodological work badly or well.

2.8. OTHER CLARIFICATIONS

Besides the principal concept of technology itself, there are other concepts that the philosophy of technology needs to clarify for its work. Among them, prominently, are the concepts of the "artifact," the "artificial," the "natural," and "nature."

a. The "Artifact"

An artifact is something that is made or used with "art" or intelligence. As such it can serve as a primary term for the implementation of intelligence in technology. Wherever there are practical artifacts, there is technology; wherever technology, there are artifacts. Artifacts range enormously in character, from stone hand axes to global communications networks, from domesticated plant and animal species to space stations, from wampum to stock markets, from witch doctor rattles to antibiotics, from cuneiform tablets to universities.

Not all artifacts need to be technological, since we have restricted the concept of technology to the useful or practical arts. There may also be aesthetic or religious artifacts. But this is of course not an area of all-or-none distinctions. The implements of the witch doctor, the alchemist, or the conjurer will be "technological" on this definition to the extent that their uses are practical, aiming at health or wealth or worldly power; they will have an additional non-technological character to the extent that other, non-practical interests are also being served. Many artifacts may be both technological and aesthetic at the same time. That a technological object—a jet airplane, for example—may be beautifully as well as efficiently designed does not make it any less technological; a goblet or a knife may be treasured both for utility and grace.

b. The "Artificial"

The adjective, "artificial," attributes to something the quality of being at least in part a product of intelligence. This certainly does not mean that what is artificial is "fake" or unreal. There is nothing unreal about an automobile or an airplane, but both are artificial. The developers of artificial intelligence in advanced computers are in quest of real machine intelligence, the product of natural human intelligence.

What may ground the suspicion of falsity that plays around the penumbra of the concept of the artificial is the fact that a bunch of artificial flowers may disappoint, be regarded as "fake," if their looks suggest that they were organic, with all the associated smells and textures that could be expected from their visual appearance. The more skillful the artifice, indeed, the stronger some people may feel a sense of disappointment and betrayal. All such examples, however, rest on the referential or semiotic (epistemological) gap between what is signified and what is actually the case. Were the artificial flowers to be considered simply in themselves, as artful handiwork in silks and satins, they would not be tarred with the suggestion of (metaphysical) incompleteness or inadequacy. They might then be appreciated for what they are, not deprecated for what they are not.

That something is "artificial" need not even mean that it is inorganic or non-living. What something fashioned with art and intelligence is made from—be it stone or silk or skin—is not the point. Thus to our earlier question (2.2.e) about the "black" tulip, it may now be responded that many living things are what they are only by virtue of the guidance of intelligence. This is most obviously true of contemporary biotechnology, in which new living forms are being created by the technology of recombinant molecular genetics—and even patented under the laws of the United States. It is also true of more traditional breeding practices of plants and animals for the deliberate achievement of certain effects in living creatures. And it is true in significant part of the landscapes into which we flee from the urban skylines.

This last point makes it important also to notice that the concept of the artificial need not be all-or-nothing. Depending on the degree to which intelligence has determined the nature of the thing, we can meaningfully speak of something as more or less artificial. The urban skyline is more artificial than the farmer's orchard; but the orchard, with its straight rows of trees and carefully cleared lanes for ease of human harvesting, is more artificial than the woods. Reforested areas are more artificial than virgin forest, but (as we noted at the outset of this book) even wilderness areas, insofar as they are the product of social decision, are to some degree artificial as well.

c. The "Natural"

The antonym of "artificial" is "natural," which attributes to something the quality of being what it is independent of the intervention of intelligence. A natural smile, a naturally pure mountain stream, a natural catastrophe are all examples of this meaning of the term. Like its antonym, the adjective can be used to express degrees of naturalness. The bulb from which the horticulturalist derives a "black" tulip is not wholly the product of intelligence, but it is not wholly what it is apart from the interventions of intelligence, either. It is to a degree artificial, but not so artificial as a silk artifact made to look like a black tulip; it is to a degree natural, but not so natural as the unbred varieties from which it has descended.

d. "Nature"

The noun "nature" is a more ambiguous term and is therefore still more perilous for clear thinking. On one use, nature is simply the collective term for *all that exists apart from the artificial.* This use, which we could call nature$_1$, is in contrast to another use, nature$_2$, that includes within its scope *all that exists in the evolving universe of space and time.* Some things, consequently, in nature$_2$, like houses and boats and bombs, are not part of nature$_1$. From this ambiguity it is easy to slip into frustrating arguments, like the one we surveyed earlier in this chapter (**2.2.d** and **2.2.e**), over whether technology is or is not "natural." Now this debate can be settled. In the sense that it is not part of nature$_1$, then technology is not natural; but it belongs to the universe of space and time and therefore is certainly part of nature$_2$. Since it is part of nature$_2$, how could it not be "natural"? The answer is that it is not "natural$_1$" but that it is "natural$_2$."

There is a third important use, nature$_3$, in which we speak of the "nature" of a thing being expressed, all other things being equal, when it *develops according to its kind without outside interference*. It is in the nature$_3$ of acorns to grow into oaks. Thus it is sometimes inferred that nature$_1$ not only has within it beings with various natures$_3$, but also that nature$_2$ as a whole has its own nature$_3$: an end or *telos*. This historically important issue, related to arguments about the existence of God, will not be pursued in this book. But another issue, whether technology is "natural" or not because, although artificial, it springs from an intelligence that is part of human nature$_3$, needs sorting out. It is true that intelligence, seen in an inherent capacity for symbolic thinking, foresight, and calculation of means, manifests itself in the human species to a preeminent degree. This means that intelligence is natural$_3$ as well as being part of nature$_2$—as much a part of nature$_2$ as clouds or claws. Thus technology, like human culture itself, can be understood as natural in both these senses, i.e., rising naturally$_3$ out of the character of *Homo sapiens*, as evolved within nature$_2$. Only when technology is considered in contrast to nature$_1$, because it is inevitably the product of artful interventions in nature$_2$ by a prominent aspect of human nature$_3$, would confusions arise.

Let us hope that these distinctions have not caused more confusions for the reader than they have cured. They would not have been necessary, perhaps, were it not true that "nature" and "natural" are words with more than purely descriptive force. They are words that also tend to bless and praise. What is "natural," we tend to feel, is approvable. "Unnatural acts," on the other hand, are offenses so abhorred by society that decent newspapers must not even name them.

Perhaps this positive connotation to "natural" comes from the normative flavor carried by nature$_3$, according to which the optimum development of the thing in question results from being allowed to develop without interference. The perfect apple comes from letting the nature$_3$ of the apple tree work itself out unhindered by drought or insects or worms. Nature$_3$, therefore, is felt to be "good." This neglects, of course, the extent to which insects and worms are busily (and, from their own point of view, constructively) working out their own natures$_3$ and the way that the weather, whatever it does to the apple tree, is part of nature$_{1,2}$.

Exactly why the word "natural" should have such a strong positive connotation is a problem too complex to be probed here. It will be enough merely to warn that technology, by intervening in and altering nature$_1$, is not necessarily evil, and by being an offspring of human nature$_3$, is not necessarily good. Perhaps with this warning and with our conceptual tools sharpened, we may move on to employ them.

CHAPTER 3

Technology and Practical Intelligence

3.1. REFLECTIONS ON INTELLIGENCE

There are of course risks involved in tying our discussion as closely to the concept of "intelligence" as is required by our adopted definition of "technologies" as *practical implementations of intelligence* (2.7). Strenuous disagreements among psychologists over approaches to precise methods of identification and measurement of intelligence warn us to be careful.

There are still more benefits to be gained, however, from making at least a general notion of intelligence basic to the philosophy of technology. It is worth surveying the ground from which we start when we begin to ask epistemological questions about the kinds of thinking that technology involves.

a. Intelligence Comes in Degrees and in a Variety of Styles

No matter how it may be defined by psychologists, intelligence will turn out to be something that allows of more and less and many domains. It is not an "all-or-none" phenomenon. The same person can exhibit more or less intelligence, e.g., at different times of day, in different areas of life—mathematical, interpersonal, artistic, mechanical, linguistic, etc.—and under differing circumstances of health, fatigue, emotion, drugs, or stress. Even the "little moron" of innumerable jokes speaks with some intelligence: just enough to make an amusing mistake. Note: the capacity to make a mistake is a sure sign of at least some degree of intelligence. Something with no intelligence at all can make no errors. The ability to be mistaken is a major intellectual accomplishment.[1]

[1]H.H. Price, *Thinking and Experience* (Cambridge, Mass.: Harvard University Press, 1953), p. 87.

b. Intelligence is not Unique to Human Beings

The continuum of intelligence extends beyond our species. Adult human beings may be preeminent (on our planet) in this trait, but it would be arbitrary to draw a sharp "intelligence boundary" between *Homo sapiens* and the rest of the animal world. Human and chimpanzee infants resemble each other in intelligence as in many other ways. Higher mammals, like apes and monkeys, horses, dogs, pigs, bears, and cats, are easily recognizable as having significant levels of intelligence. Similar in this trait, we now find, are aquatic mammals like whales and dolphins. The notion of intelligence is firmly grounded within the organic order.

c. Intelligence Relates to the Capacity for Flexible Response

We withhold the adjectives of intelligence from routines of behavior, however complex and otherwise impressive, that are fixed and unresponsive to changes in the environment. If a behavior is generated internally, "by instinct" (as it may roughly be called), so that it is "programmed" or "hard-wired" into the organism's repertoire, incapable of major modification under changing circumstances, it is not classed as intelligent behavior. Similarly, in humans, behavior that is a simple reflex (like the notorious "knee-jerk" reaction) or that is random and unresponsive to the environment (like an epileptic seizure or like some drunken behavior) is to that extent excluded from the scope of intelligence. Conversely, the greater the flexibility of appropriate response to changing circumstances, the greater the degree of intelligence.

d. Intelligence Varies with Speed of Response

There is a time element implicit in the notion of intelligence, represented in our language by the colloquial synonym "quick-witted" for "intelligent," and by the gentle euphemism "slow" to replace "less intelligent." When changing circumstances are rapidly answered by appropriate responses, we acknowledge a higher degree of intelligence than when routines are altered only after much time.

e. Intelligence Varies with Fineness of Discrimination

The capacity to make mental distinctions, whether in complex environmental circumstances, in aesthetic appreciation, or in complex linguistic domains, is generally recognized as a mark of intelligence. If only gross discriminations can be made, e.g., between predator ("It eats me.") and prey ("I eat it."), some degree of intelligence may still be recognized; but as discriminations become finer or more subtle, the level of intelligence manifested is higher.

f. Intelligence Varies with Remoteness of Inference

Intelligence is expressed in the capacity to anticipate what is not here and now by means of signs that point to events at some remove in time or space. When the mouse reads the body language of the cat and successfully dodges away from the cat's pounce, there is some intelligence in that performance, though the sign (the crouch) came almost immediately before the signified event (the spring). There is still more intelligence involved in the rat's recognizing the proper turns to make in finding the right way through a maze to the more remote prize at the end. On the human level, there is intelligence at work in a game of chess when one can anticipate the next move of the opponent; there is still more intelligence demanded for anticipating an opponent's strategy at several removes. The longer the inference chain, in general, the higher the estimation of the intelligence involved.

g. Intelligence Varies with Synoptic Power

When one or two well-discriminated signs are put together to make an inference, intelligence is at work; but still more intelligence is shown when several or many signs are successfully pulled together into a meaningful whole. Not only quickness, not only clarity of discrimination, not only remoteness of inference, but also grasp of complexity into unity is a measure of intelligence. When available signs give conflicting indications, then priorities among the signs need to be set, normal inference patterns need to be questioned, complex resolutions achieved. The more varied the data given to be handled, the greater the unifying intelligence required.

h. Intelligence Varies with Effectiveness in Achieving Appropriate Goals

Even quick discriminations of complex circumstances would hardly be classed as intelligent were these not tied to appropriate outcomes for the organism. This roots the notion of intelligence firmly in the domain of purposes, intentions, aims, and norms. Much discussion in the philosophy of physical,[2] social,[3] and biological[4] science has revolved around the concept of purpose or "teleology" in scientific explanation. There is considerable reluctance among many philosophers (especially among those who find physics to be "basic" in the honorific sense that all other sciences and all human knowledge should ultimately be expressed in its concepts alone) to acknowledge purpose as an essential category. Living, goal-seeking organisms are an embarrassment to the project of accounting for all happenings while using only explanatory concepts and laws that are suited to the non-living, non-purposive processes studied by physicists. But if philosophers, committed to comprehensiveness in thinking, are to take the existence of physicists and their intelligent activities into account as well as the physical world that physicists study, then the

[2]Carl G. Hempel, *Philosophy of Natural Science* (Englewood Cliffs, N.J.: Prentice-Hall, 1966), especially Chaps. 6, 7, and 8.

[3]Richard S. Rudner, *Philosophy of Social Science* (Englewood Cliffs, N.J.: Prentice-Hall, 1966), especially Chap. 5.

[4]David Hull, *Philosophy of Biological Science* (Englewood Cliffs, N.J.: Prentice-Hall, 1974), especially Chap. 4.

realm of purposes, ends, and aims is unavoidable. Wherever intelligence is present, those "strange objects," described by molecular biologist Jacques Monod as manifesting "telenomic"[5] traits, cannot be avoided. Since flexible responses aimed at achieving appropriate goals abound in the organic world and human society, the attempt to avoid teleological categories in the philosophy of technology seems clearly unprofitable.

i. Intelligence Relates to the Appropriateness of Goals Themselves

Not only responding with flexibility, speed, and discrimination toward effective means, but also aiming at optimal goals, is an important measure of intelligence. Ends, as well as the means to those ends, are always open to assessment. All goals are not equally intelligent. When ends are "fixed," beyond criticism, we may suspect compulsiveness or fanaticism, though some ends may turn out not to be optional. How one might approach the intelligent assessment of goals ("intrinsic goods"), as well as the means to these goals ("extrinsic goods"), is a topic in ethical theory that will occupy us later (6.1).

3.2. PRACTICAL *VERSUS* THEORETICAL INTELLIGENCE

It must have occurred to the reader that the listed tasks typically associated with intelligence are immensely diverse. One traditional way of finding order within all this diversity is by emphasizing the distinction between *practical* and *theoretical* intelligence (or "reason" as it is sometimes called[6]).

We see practical intelligence in action when we watch a rat learn the way through a maze; we acknowledge theoretical intelligence when we follow someone's elegant mathematical proof. What do these forms of intelligence have in common, and what are their differences?

One common feature is that both are *purposive*. But this does not get us far: the purposes seem profoundly unlike. The general purpose of practical intelligence is to survive or thrive; the general purpose of theoretical intelligence is to know or understand. Other basic similarities consist in the extent to which both practical and theoretical intelligence are found in *degrees of greater and less:* degrees measured by quickness, power of discrimination, remoteness of inference, and capacity for integration. Practical intelligence uses its available speed to avoid danger or take advantage of changing circumstances; theoretical intelligence uses its speed to unravel a conceptual obscurity. Practical intelligence discriminates subtle differences

---

[5] Jacques Monod, *Chance and Necessity: An Essay on the Natural Philosophy of Modern Biology* (New York: Vintage Books, 1972), pp. 13–22.

[6] Immanuel Kant worked out the distinction in great detail in his masterful *Critique of Pure Reason* (1781, 2nd ed. 1787) and *Critique of Practical Reason* (1788). There are advantages, however, to shifting our term to "intelligence" from "reason," if only because "reason" has acquired so many overtones and lofty associations in the long literature surrounding it.

in the environment to plan more successful actions; theoretical intelligence discriminates subtle differences among ideas to provide itself with better analyzed premises. Practical intelligence infers remote consequences to prepare for or avoid events; theoretical intelligence infers remote conclusions to follow the argument where it leads. Practical intelligence synthesizes its data to provide a unified battle plan for life; theoretical intelligence synthesizes to provide wholeness of understanding.

These differences of purpose and working are not simply different. They seem actually to conflict, at least much of the time. Consider the following conflicts:

(a) Practical intelligence is inevitably involved with the surroundings and is always interested in some outcome. Theoretical intelligence, on the contrary, does best when disinterested or impartial. Theory is in danger of distortion and falling into mere rationalization if it becomes involved in the passions and partisanships of daily affairs.

(b) Practical intelligence cares about what works; and if its solutions work as reliably as desired, it is satisfied. Theoretical intelligence cares about what is really so, as extensively as can be determined, and is not at all satisfied with merely successful procedures. If a successful procedure is based on a poor model of how things are, theoretical intelligence takes no pleasure in it. On the contrary, theoretical intelligence demands to understand why a procedure should be successful, if it is. The solutions of practical intelligence become the unsolved problems of theoretical intelligence. One wants to stop and enjoy the benefits, while the other insists on criticizing and pushing on.

(c) Practical intelligence judges ideas by their applicability to the task at hand and by the degree of reliability of the result needed. This means that the simpler the ideas, supposing that they work well enough for the task as defined, the better they are for practical purposes. Practical intelligence must reject needless complications as temptations to inefficiency. Theoretical intelligence, on the other hand, must reject the lure of simplicity if ease of application gets in the way of depth and precision.

(d) A similar conflict appears in the standards used by practical and theoretical intelligence for success. When needs of practice are the governing aim, success is judged by the net effect of one's ideas in action. What counts—and all that counts—are the consequences. The success of theoretical intelligence, however, must be judged by the quality of the path taken. If the rules protecting sound theory have been broken, the short cut spoils the game. The success of theoretical intelligence cannot be assessed apart from its own principled norms.

(e) Highly intelligent activities, like flying an airplane, may be done better without excessive explicitness of the principles embodied in the activity. If a pilot, for example, while correcting for gusts on the final approach to the runway, is too much taken up with calculating finer points in the laws of aerodynamics, the landing is likely to be less than smooth. For theoretical intelligence, however, there seems to be no limit to the benefits of explicitness. Theoretical intelligence does not begin with explicitness, but it presses patiently toward it. The nature of theoretical activity assumes a kind of freedom from pressures, in principle, that practical intelligence cannot enjoy. And in this theoretical context it is better that every assumption, every premise, every inference be drawn out of the background and made as explicit as possible. Only when ideas become explicit can we winnow out the weak from the strong and correct the weak. "Tacit," seat-of-the-pants, knowing is often vital to intelligent *performance*.[7] but theoretical intelligence, though inarticulate at the start, seems to have a built-in need to bring the global, tacit dimensions ever more focally into view—struggling not to lose richness through the processes of linguistic abstraction—for the sake of critical *understanding*.

(f) The sorts of value that can be claimed by practical and theoretical intelligence, finally, seem to conflict. Practical intelligence always has instrumental value, on the one hand, as a means or instrument for the achievement of other valued ends. Theoretical intelligence, on the other hand, the thirst for understanding for its own sake through the playful exercise of unfettered mentality, rises for at least some persons to the status of an intrinsic value. It has its own aesthetic as well as ethical imperatives. Sometimes its claims for respect and resources seem to cut athwart the urgent practical needs of the time. Then the differences between practical and theoretical intelligence strike the hottest sparks. As Whitehead wrote, "There is a strong moral intuition that speculative understanding for its own sake is one of the ultimate elements in the good life. The passionate claim for freedom of thought is based upon it. Unlike some other moral feelings, this intuition is not widespread. Throughout the generality of mankind it flickers with very feeble intensity."[8]

There are many grounds, then, to deal separately with practical and theoretical intelligence. They are both expressions of mentality, but they each headed separate households before their recent marriage in modern technology and science.

3.3. TRADITION-BASED PRACTICAL INTELLIGENCE

In a memorable discussion, Whitehead names practical intelligence the "Reason of Ulysses."[9] Ulysses (or Odysseus), the great mythic hero of Homer's *Odyssey*, was above all the master of managing to get himself and his crew out of scrapes as they sailed the "wine dark seas" of their adventures. Whether trapped in the cave of the Cyclops or needing to slip past the irresistible sirens, Ulysses always came up with a clever method to deal with his practical problem of the moment.

In this fabled capacity to envision and implement a novel solution for a pressing challenge, Whitehead finds a paradigm for practical reason, the biologically deeper and evolutionarily older branch of intelligence, the aspect of intelligence that we

---

[7]Michael Polanyi, *Personal Knowledge* (Chicago, Ill.: The University of Chicago Press, 1958, 1962). Also his *The Tacit Dimension* (Garden City, N.Y.: Doubleday, 1966).

[8]Alfred North Whitehead, *The Function of Reason* (Boston, Mass.: Beacon Press, 1929), p. 38. Reprinted by permission of Princeton University Press.

[9]*Ibid.*, p. 10.

share "with the foxes."[10] What this involves, concretely, is the capacity to take account of the physically absent[11] for the sake of serving the appetite of living things to live and to improve the conditions of their lives. *Mentality* is at root *the capacity to envisage possibilities that are simply "other" than the actualities of the immediate environment. Practical intelligence*, on the other hand, is *the capacity for mental self-discipline in the service of the urge of life.* As Whitehead puts it: "Reason is a factor in experience which directs and criticizes the urge towards the attainment of an end realized in imagination but not in fact."[12]

Mentality alone, in the service of life's appetite "to live, to live well, to live better,"[13] can be dangerously anarchical. Possibilities abound. Which one to choose? Whitehead imagines an extremely low level of mentality posed by a terrible thirst in the desert. The thirsting traveler can think of nothing but water, although there is no water in the immediate environment. Such a "slavish" mentality, enslaved by negative appetition from what is present, could kill the organism if it is not self-disciplined. Running at random here and there, looking for water first behind this sand dune and then beyond that cliff in the frantic but unorganized pursuit of pure possibilities, will soon destroy itself. But if mentality can criticize its own appetitions and discriminate among its own imaginings, if it can bring order into its search for the means of life, then mentality can be put to powerful service in the art of life. The Reason of Ulysses, practical intelligence, is the self-regulative capacity of mentality. "It introduces a higher appetition which discriminates among its own anarchic productions."[14]

What practical intelligence of this sort primarily provides to mentality is *method.* A method is a regular way of achieving an abstractly envisaged aim. A fortunate method embodies a mental achievement and perpetuates it. If the thirty traveler in the desert finds enough self-discipline to lie in the shade of an outcropping during the heat of the day, to travel only at night, then to walk only in a straight line guided by the North Star, and perhaps to dig for and chew the roots of juicy plants—to impose method on mentality—there is a greater chance of survival. And if this method works once, it will commend itself to practical rationality as a possible way of escaping desert thirst another day under similar circumstances.

Practical intelligence, as a mental phenomenon, deals with ideal possibilities. Motivated by the urge to live and to thrive, practical intelligence sorts these envisaged possibilities into orders of relevance for realization and attempts to guide action into the fruitful channels of regular method. Once a fortunate method is found to serve the urge to live and thrive, it may be remembered and retained as a complex new ideal possibility, and when a recognizable need arises, intelligence

[10]*Ibid.*, p. 10.

[11]Price, *Thinking and Experience*, p. 34.

[12]Whitehead, *The Function of Reason*, p. 8.

[13]*Ibid.*, p. 8.

[14]*Ibid.*, p. 34.

may make it available again for action. This may happen with an individual organism. Then we say that the organism has "learned" a new method. Still more importantly, this may happen socially, when others learn and adopt a method, passing it on from generation to generation. Then we speak of a "culture" resting on shared methods that are transmitted by teaching and learning.

Human culture is the domain on which this book is focused and is co-extensive with human existence from its earliest traces. There are today no known human communities without learned methodologies of some sort, and the identification of human presence in the past is often made by paleontologists and anthropologists simply by locating evidences, like artifacts, of cultural life.

An artifact is a way of implementing a methodology. Finding an arrowhead is discovering a learned method of acquiring food by hunting. Locating a hoe is discovering another method of acquiring food by agriculture. Human beings from time immemorial have used the Reason of Ulysses to support the urge to live and to thrive by implementing envisaged practical possibilities, i.e., by technologies.

The history of human technologies shows that after a discovery is made of some method with its associated artifacts, the new complex is quickly brought to maturity within its own terms, is radiated geographically, and then is repeated and repeated by the adopting cultures. This is the repetition of tradition, through which successful methodologies, the precious past achievements of practical intelligence, are safeguarded.

We owe much to tradition, since the vastly greater part of human history has relied on it for all aspects of culture, including the technological. Tradition is highly conservative of methods that have worked. Learning is done primarily by imitation: like mother, like daughter; master and apprentice; the secrets of the guild. Immense achievements have been based on such tradition-based technologies, from the Roman aqueduct system to the lofty Gothic cathedral, and from the smelting of iron to the building of wooden sailing ships and navigating them by magnetic compass.

The long history of practical intelligence embodied in culture and perpetuated by tradition allowed for gradual improvement of methods, since the urge of life to thrive—"to live, to live well, to live better"—continues to motivate within the framework of tradition. But the primary thrust of practical intelligence, once it has achieved a method that works, is to protect and repeat it. "If it isn't broken, don't fix it," is the authentic voice of practical wisdom.

There is a tendency in practical intelligence to show not only indifference but even hostility to the criticism of effective methods. This is understandable because of the frequency with which the anarchical tendencies of mentality, undisciplined by reason, tend not only to fail but even to threaten the hard-won achievements of the past. If one's ancestors have successfully built the ceilings of innumerable stone cathedrals in a certain way, it may be reasonable to reject new suggestions with some impatience. This tendency toward impatience with criticism of established methods Whitehead calls "obscurantism":

Obscurantism is the inertial resistance of the practical Reason, with its millions of years behind it, to the interference with its fixed methods arising from recent habits of speculation. This obscurantism is rooted in human nature more deeply than any particular subject of interest. . . . Obscurantism is the refusal to speculate freely on the limitations of traditional methods. It is more than that: it is the negation of the importance of such speculation, the insistence on incidental dangers.[15]

Although obscurantism is the negative side of successful methodology, it would perhaps be better to accept obscurantism than the anarchical destruction of past mental achievements—if only tradition *could* in fact fully preserve the discoveries of the past. The law of entropy, however, i.e., the law that disorder in closed systems tends to increase,[16] holds not only for thermal systems but also for cultures. The same principle, e.g., that my car battery will always tend lose its charge, when left alone, unless I generate more energy to keep it up, describes the way that a tradition-based methodology, left alone, will gradually lose its quality. Information, once lost to a system, is never recovered without fresh efforts at rediscovery. In the long history of traditional human technology, practical intelligence has been both the source of new information *and* the inhibition against change. It is this ambivalence within practical intelligence that may help us understand both the rise of technology over the ages and its slow pace of change over the thousands of years that make up the vast bulk of human history.

3.4. THEORY-BASED PRACTICAL INTELLIGENCE

Something happened to the pace of technological change in Europe and America about two centuries ago. Practical intelligence became closely linked to theoretical intelligence for the first time in human history. In virtually no time, as prior ages would judge, the world has seen an unprecedented explosion of inventions, artifacts, and devices that have altered not only large domains of human culture but also the physical world in profoundly significant ways.

It is easy, reflecting on the many differences—and even conflicts—between practical and theoretical reason, to see why these two forms of thinking might have been out of contact for so long. The harder problem to understand is why at some point they should ever have come fruitfully together. And why should this event have happened in Western Europe just when it did? These are intriguing questions in their own right, and surely belong to the philosophy of technology to consider; but we must set them aside here. The more immediate question is what happens to the Reason of Ulysses when the characteristic traits of theoretical intelligence are brought effectively into practical thinking.

In 3.2, above, we looked at these traits. Theoretical intelligence tends, in contrast to practical intelligence, to be (a) *disinterested*, (b) *curious* about what is so

---

[15]Whitehead, *The Function of Reason*, p. 43.
[16]*Ibid.*, pp. 23–24.

and why methods work, if they do, (c) concerned for *detail*, (d) highly *self-critical*, and (e) *explicit.* Let us look at what this has meant when linked to the urge "to live, to live well, to live better."

(a) The Reason of Ulysses is always keenly interested in the circumstances and the outcome. This is the motivating strength of practical intelligence in solving immediate problems, if it can; it is also a besetting weakness if it leads to short-sightedness or self-delusion. There is nothing impartial about Ulysses. He does not ask what is fair to the Cyclops or what precedents for the long run he may be setting with Circe. If there is to be a corrective breadth of view for long-run policies, it will have to be supplied by the coolness of theoretical intelligence, which in this way places the partial self-discipline of practical thinking under still further discipline.

(b) Practical intelligence is only concerned *that* a method works. When it works well enough for the purposes at hand, practical intelligence is satisfied. Theoretical intelligence, on the other hand, demands to know *why* a method works. What is behind the success? What is the world like, so that it behaves as it does? When theoretical understanding is achieved, partial successes may be strengthened and made more effective because theoretical intelligence envisages new possibilities for corrected methods. More, the local successes of practical intelligence may be able to be generalized into new areas of practice. Under the leadership of mental excursions into the possible from the actual, whole new areas of possible *wants* may be opened up, once the structure of things becomes known in theory. The deliberate pursuit of inventions begins. "If that's the way things are . . . ," muses theoretical intelligence, ". . . then *this* ought to work," completes practical thinking, and starts an invention—not (as in traditional culture) to solve a pressing practical problem but to gain an abstractly conceived improvement that might otherwise never even have been desired.

(c) Practical intelligence cares only for enough detail and precision to get the job done as effectively as needed. Many methods are "good enough for government work," though not particularly deep or thorough. Theoretical intelligence, on the other hand, is thirsty for *exact* knowledge. Nothing less will do, since its acquisition is the end itself, not a means to something else that allows a rough approximation to "do." "Good enough" is not good enough where understanding for its own sake is the prize. The pressure for precision and detail from theoretical reason, once linked to practice, forces practical intelligence to attend to ever closer tolerances. Only when linked to an interest in ever-increasing precision are machine tools, for example, possible. The whole infrastructure of modern technology rests on precise measurements and exact tolerances. Theoretical intelligence, as guide and goal, holds up the standard to be met. It is perhaps an impossible standard, in practice, since the abstract ideals of theory demand infinite detail and complete perfection. But as a demanding standard it has been immensely important to further the disciplining of the Reason of Ulysses.

(d) Practical intelligence knows no other standard than success. "If you're so smart, why aren't you rich?" is the taunt of the practitioner against the theorist.

The reply of theoretical reason must be that riches are no guarantee of truth or understanding. Wrong ideas are often associated with practical success. Roman generals sometimes based their tactics on auguries obtained from the entrails of birds—and won their battles anyway. Theoretical intelligence will not still its curiosity or blunt its criticism merely because of the practical successes of traditional methods. This means that the long, drowsy periods of rote learning and technological repetition characteristic of traditional cultures will be denied to technologies based on theory. Theoretical intelligence serves as a gadfly to sting away the heavy sleep of obscurantism. That is the surest way to continual change.

(e) Practical intelligence, finally, is often "tacit" about its premises and assumptions. This means, as we saw, that when its proposals fail there is no clear way to diagnose the reasons for the failure. Expert intuition may often be highly effective in bringing intelligent purpose to bear on presented circumstances; but if even the expert must remain inarticulate about *reasons* for actions taken, correction or improvement becomes virtually impossible. Theoretical intelligence, on the other hand, demands explicitness to the greatest possible extent. When a technology is based on theory, and when theory is well articulated, then individual assumptions may be isolated and checked for their contribution to the total effect. Just as a good car mechanic isolates the problem before digging into the engine, so a critic of theory-based technology can examine and modify elements systematically.

Not only is this a far more effective method of correcting for failure than previous all-or-nothing approaches, it represents for the first time in history a method for the systematic and deliberate pursuit of new inventions. Mental envisionment ("What do I want?"), articulation of theoretical consequences ("What would happen if . . . ?"), construction of an artifact ("Will this do it?"), empirical observation of some actual outcome ("Did it work?"), comparison ("Exactly what went wrong?"), rearticulation of theory ("Perhaps if instead of . . . "), isolation of elements ("Was *this* the problem?"), modification of the artifact ("*Now* let's see"), fresh empirical observation ("That's better, but . . .")—this cycle between theory and practice constitutes "the invention of the method of invention."[17] In these ways, the characteristic contributions of theoretical intelligence to technology have changed the world.

[17]Alfred North Whitehead, *Science and the Modern World* (New York: The Free Press, 1929, paperback edition, 1967), p. 96. Copyright © 1925 by Macmillan Publishing Company, renewed 1953 by Evelyn Whitehead. Reprinted with permission of the publisher.

CHAPTER 4

Technology and Theoretical Intelligence

Toward the end of the previous chapter we looked at theoretical intelligence as though it were something that could "supply" and "demand" and "reply" in various ways. That may be a useful figurative way of speaking, but "it" is not a thing. It is more accurately considered a second-order function of mentality: a function with a special character and goal. Whitehead calls it the "Reason of Plato."[1]

4.1. UNIMPLEMENTED THEORETICAL INTELLIGENCE

Plato will be our paradigm for theorists, just as Ulysses has been our paradigm for escape artists. Ulysses, as we have seen, shares his practical intelligence "with the foxes," but Plato shares his theoretical perspective "with the Gods."[2] There is a sense in which theoretical intelligence tries to survey the whole world from "above." In this sense practical intelligence becomes an item functioning inside the world while theoretical intelligence tries to transcend everything in the world in its aim to understand it. The paradoxes of "comprehensiveness" (1.3 and 1.7) reappear as theoretical reason attempts to understand even itself.

As far as we now know, the Reason of Plato is not found outside the human species. Its functioning depends on the high development of an abstract symbolic system and the capacity to manipulate symbols in sophisticated ways, far removed from local circumstances. Even among the human species, it may not be evenly dis-

---

[1]Alfred North Whitehead, *The Function of Reason* (Boston, Mass.: Beacon Press, 1929), p. 10.
[2]*Ibid.*, p. 10.

tributed, and it is, compared to practical intelligence, a recent development on our planet. Whitehead dates its first significant appearance at roughly 6000 years ago, along with the beginnings of civilization.[3] It may have functioned earlier, if cave paintings and religious objects dating from the pre-historic past suggest that mythic imagery functioned in human thinking as attempts at understanding the world, not merely at magically controlling it.

However this may be, the self-discipline of theoretical intelligence by its invention of logic—mental controls against the "anarchical" tendencies of speculation—dates within the historic period, perhaps less than 3000 years ago. We credit the ancient Greeks with inventing logic, and simultaneously discovering that sheer conceptual play can itself be controlled by method. That means that it is possible to *argue* and not merely *proclaim*. To the *comprehensive* weaving of envisioned possibilities were added the powers of *critical* evaluation. This is the Reason of Plato, or theoretical intelligence, as the term will be used here. First let us consider the strengths and weaknesses of such intelligence apart from its implementation through technology.

(a) A great strength of theoretical intelligence is its profound *claim to intrinsic value* (3.2.f), on the basis of which it need not be justified by any other achievement or result. This intuition of its having value for its own sake, shared by at least some people at virtually all times in the last three millennia, has assured its continuation and development, whatever its obstacles.

Plato's teacher and hero, Socrates, was one who strongly embodied the intuition of the intrinsic value of theoretical intelligence in his own life as well as in his teaching. After a long career of making himself unpopular with the obscurantists of his day as the gadfly of theoretical reason, Socrates was condemned to death, allegedly for corrupting the young with his questioning ways and his unorthodox views. Awaiting the time of execution in his cell, Socrates was offered an opportunity to flee into exile. Among other reasons for gently but firmly rejecting the offer, Socrates points out that at his trial he had argued that "an unexamined life is not worth living."[4] If even "enlightened" Athenians could not tolerate his ventures in theoretical intelligence, even less could the uncultivated lands of exile be expected to allow him to continue his examinings.[5] Death would be better than the abandonment of his life's mission.

The passion of reason for its own dispassionate pursuit has erupted many times in history since Socrates. An especially poignant element in the struggle between Galileo and the authorities of his day was the sense of the intrinsic importance of unfettered thought. Despite threats of torture, public humiliations, and house arrest,

---

[3]*Ibid.*, p. 40.

[4]Plato, *Euthyphro, Apology, Crito*, trans. F.J. Church, rev. trans. R.D. Cumming (Indianapolis, Ind.: Bobbs-Merrill, second edition, 1956), p. 45.

[5]*Ibid.*, p. 64.

Galileo continued to think his theories, and the world of thought was changed.

(b) Another strength—which can also be a weakness—of theoretical intelligence is its readiness to lavish endless care on the *discrimination of detailed distinctions* (3.2.c and 3.2.e). Socrates inspired Plato, who taught Aristotle the importance of careful distinctions. Centuries later the Aristotelian tradition of discriminating discriminable ideas was in full flower among the scholastics of the Middle Ages. The opponents of Galileo were not irrationalists; they were solidly grounded in the vast achievements of theoretical intelligence gained by their predecessors. The revolt of Galileo was not against unreason but against what Whitehead calls the "inflexible rationality of Medieval thought."[6]

With the entrance of *method* into theoretical intelligence comes the inevitable temptation to obscurantism, the excessive attachment to a sometimes-successful method. The method of careful discriminations may be illuminating—or it may entangle thinking in a morass. Much depends on the sorts of discriminations and distinctions that are made.

(c) Theoretical intelligence distinguishes itself for *remoteness of inference* (3.1.f). Once the controls of logic were discovered and employed, chains of argument could be linked nearly to infinity. The earliest pre-Socratic scientist-philosophers thought their way all the way from daily experience to the fundamental *archē* from which everything was made. Aristotle argued all the way from the character of change in the natural world to the necessity of a God.

(d) In forging such long chains of inference, the Reason of Plato seeks to bind understanding into a comprehensive *unity of synoptic vision* (3.1.g). The pre-Socratics had an amazing power of synthesis, leaping to unifying theories of reality that remain important today. The main theoretical alternatives about reality were mapped out: is the universe a continuum with local deformations (as contemporary field theory might suggest and as Heraclitus argued over a century before Plato), or is it finally granular with clumpings and aggregations of various sorts (as contemporary particle theory might suggest and as Democritus proposed)? The free power of untrammeled theoretical intelligence to unify thought into systems is one of its glories.

(e) The limits of the Reason of Plato are not found in its capacity to discriminate and to synthesize but in its incapacity to give a *basis for choice* among the many fascinating pure possibilities that it can generate. Seen as brilliant speculations with intrinsic value, there is no reason to choose. But seen as mutually incompatible models claiming to be true depictions of the universe, there is a need for some principle of elimination—or at least some method for narrowing down the proliferation of possibilities for those who would believe them.

---

[6]Alfred North Whitehead, *Science and the Modern World* (New York: The Free Press, 1929, paperback edition, 1967), p. 8.

4.2. IMPLEMENTED THEORETICAL INTELLIGENCE

Some readers may have wondered why there have to this point been so few uses of the word "science." This has been no accident. It is too easy to say that "science" is the secret of the vast development of technology in the modern world. It is also too simple to say that technology is the key to modern science. Both statements are partially true and partially false. Modern science, as it has developed since the seventeenth century, is the joint product of theoretical and practical intelligence, and so is modern technology. Neither gave birth to the other. They are non-identical twins of the same parents.

a. Outfitting Plato with the Tools of Ulysses

Modern science would be unthinkable without its embodiment in instruments and apparatus. Pre-modern science, as mainly unimplemented theoretical intelligence, was brilliantly rational; Galileo Galilei (1564–1642), the father of modern science, fought such naked rationality with theories and inferences linked closely to artifacts and observations.

The pendulum and the inclined plane, as artifacts designed to restrain or slow down the motion of a falling object, so that it could be better studied, helped Galileo revolutionize physical dynamics; but even more dramatic is the part played by the glass lens in the transformation of human thinking about the astronomical universe. The traditional craft of glass making is ancient, and the discovery of the optical properties of irregularities in glass is old, as well. Eye glasses date from the thirteenth century. The Arab astronomers, still earlier, had used long tubes to view the stars with care. In 1605 the Dutch optician, Johann Lippersheim, put lenses into a viewing tube and thus invented the telescope. On hearing the news in 1609, Galileo almost immediately created one for himself and turned it toward the sky at night, amazing himself with the discovery of many more stars than had been previously imagined and with the view of what now looked like mountains and craters on the moon.

It is important to remember that Galileo did not look through his telescope without having theory in his mind. It was the Reason of Plato that directed the artifact made by the Reason of Ulysses. There were in fact two theories available to Galileo, one of which was strongly preferred by him but so far without a way of excluding the other. Both were brilliant examples of theoretical intelligence.

One, the achievement of the great Alexandrian astronomer, Ptolemy (2nd century, A.D.), was the detailed mathematical culmination of principles laid down by Plato and Aristotle, accounting for the motions of the stars and planets by assuming the centrality of the earth. All heavenly motions, as Plato had required,[7] were described as uniform motions in perfect circles, "worthy" of the divine movers themselves. Even the retrograde and apparently irregular motions of some of the

---

[7]Plato, *Timaeus*, trans. F.M. Cornford (Indianapolis, Ind.: Bobbs-Merrill, 1959), p. 27.

planets were understood as uniform circular motions, with centers located in other uniformly moving circles. These "epicycles," when added to the larger astronomical cycles, made the total resultant description of planetary motions square reasonably well with the observations collected. As discrepancies were noticed and became too great for intellectual comfort, additional epicycles could be added to Ptolemy's theory to make the system mentally satisfying again.

The other, competing, achievement of theory, anticipated in ancient times by Aristarchus of Samos (second century B.C.) and preferred by Galileo even before he picked up his telescope, was an intellectual creation by Copernicus (1473-1543), who envisaged the possibility of an astronomical universe with a central sun. Copernicus was profoundly devoted to the honor and majesty of the sun, and considered his postulating its centrality to be much more appropriate than relegating it to being merely another of the planets, as in Ptolemaic theory. In *De Revolutionibus* Copernicus wrote glowingly of his alternative order:

In the middle of all sits Sun enthroned. In this most beautiful temple could we place this luminary in any better position from which he can illuminate the whole at once? He is rightly called the Lamp, the Mind, the Ruler of the Universe; Hermes Trismegistus names him the Visible God, Sophocles' Electra calls him the All-seeing. So the Sun sits as upon a royal throne ruling his children the planets which circle around him. The Earth has the Moon at her service. . . . Meanwhile the Earth conceives by the Sun, and becomes pregnant with an annual rebirth.[8]

Perfect circular motion and epicycles were still accepted as theoretical necessities, however, by Copernicus. Even the number and complexity of epicycles were roughly the same on his theory as on Ptolemy's. As Thomas Kuhn notes, "Judged on purely practical grounds, Copernicus' new planetary system was a failure; it was neither more accurate nor significantly simpler than its Ptolemaic predecessors."[9] Unimplemented theoretical reason, down to the time of Galileo, was (again) experiencing difficulties in offering more than aesthetic or qualitative grounds for choice between brilliant speculations.

Theoretical intelligence with a telescope to look through, however, is in a vastly different position. Galileo started with a practical artifact used in the context of a theory; then came an observation of the difference made by the intervention of the artifact, and repeated observations (on the second night Galileo found that the shadows on the moon's surface had shifted enough to make the lunar contours even more clearly the contours of mountains); then followed contemplation of the meaning of these observations and a further inference to more remote theoretical

[8]Nicholas Copernicus, *De Revolutionibus Orbium Caelestium* (1543), First Book, chap. 10, trans. John F. Dobson and Selig Brodetsky in *Occasional Notes of the Royal Astronomical Society* (London: Royal Astronomical Society, 1947), vol. 2, no. 10. By permission of the Royal Astronomical Society, and quoted from Thomas S. Kuhn, *The Copernican Revolution: Planetary Astronomy in the Development of Western Thought* (Cambridge, Mass.: Harvard University Press, 1966), p. 177.

[9]Kuhn, *The Copernican Revolution*, p. 171.

consequences (would other planets, like Jupiter, show similar features, and would the "inner" planets show phases like the moon?); then more theory-controlled observation, including the discovery of the four principal moons of Jupiter and the phases of Venus; from this resulted the strengthening of Copernicus' theory-anticipated analogies, between the earth and Jupiter as both planets circled by moons, and to the closed orbit of Venus around the sun, not the earth. Characteristically modern scientific astronomy was born not with Copernicus' brilliant intellectual leap, but with Galileo's thorough blending of the Reason of Plato with that of Ulysses.

A similar blending was achieved in Italy shortly after, by Galileo's younger colleague Torricelli (1608–47) in his discovery of air pressure—or was it the invention of the barometer? The craft of the glass makers, especially the advanced techniques of the guilds of Venice, once again led the way. Without their tradition-based skills, Torricelli could not have tested the theory (suggested by Galileo's general idea that air has weight and the puzzling inability of water pumps to raise columns of water higher than 34 feet) that the atmosphere at the surface of the earth has enough weight, pressed down by a great invisible ocean of air above us, to push liquids, depending on their weight, a certain distance up into empty tubes. Led by this model, Torricelli designed his artifacts, a transparent glass tube full of mercury (open at one end) and a container of mercury, and tested the height to which the mercury would fall when the open end was inverted into the container. Since mercury weighs about 14 times water, the height of the column of mercury visible through the glass tube should turn out to be 1/14th of 34 feet, and so it was! Another consequence of the theory would be that less air would press down on the surface of an open dish of mercury at the top of a high mountain than at its foot. This inference led to experiment under the direction of Pascal (1623–1662) and to the famous expedition up the Puy de Dôme to gather theory-directed observations.[10] Sure enough, the length of the column of mercury, visible through the glass tube, that could be balanced by the weight of air at a high altitude was less than had been measured below. An instrument to measure air pressure and the confirmation of the concept of air pressure were achieved at the same time.

These two examples from the early history of modern science, which stress the essential role of implemented intelligence in modern science from its beginning, could be multiplied almost infinitely. The vital role of sophisticated instrumentation and apparatus today is even more obvious.

b. **Expanding Ulysses' Vision with the Perspective of Plato**

The blending of theoretical and practical intelligence works both ways. Just as modern science is unthinkable without its artifacts, so modern technology is un-

[10] A.C. Crombie, *Medieval and Early Modern Science*, Vol. 2, *Science in the Later Middle Ages and Early Modern Times: XIII-XVII Centuries* (Garden City, N.Y.: Doubleday Anchor Books, 1959), p. 250.

imaginable without its ingredients of theory. This is above all striking in what is now called "high technology," based on esoteric advances in chemistry, biology, and physics. Since those technologies will occupy us for most of the remainder of this book, they may for now be set aside. Even the "paleotechnic"[11] steam engine will serve for an example.

The first known apparatus for pumping water with atmospheric pressure is called "Hero's Ball" or "Hero's Fountain" in honor of Hero of Alexandria (first century, A.D.). Hero was both a mathematician and an inventor of his time, but the Reason of Plato and the Reason of Ulysses, though present in one or another of his works, were never blended. Hero's Fountain was a closed container, partially full of water, into which air could be blown to build up pressure. The pressure in the enclosed vessel, when released, would push the water up through a pipe to squirt merrily in the air for the amusement of court guests. Hero's Ball, on the other hand, is best exemplified today by the (somewhat outmoded) siphon bottle at cocktail parties. Both exhibit primitive applications of the principles of pneumatics and hydraulics, about which Hero himself wrote in texts that were influential in Islamic as well as European circles for the centuries to come.

It was not until these applications were systematically generalized and refined in eighteenth century Europe, however, notably by Thomas Newcomen (1663-1729) and later even more notably by James Watt (1736-1819), that modern industrial technology got its start. The Reason of Ulysses, searching for some novel method to pump water out of mines so that coal, the new energy source, could be more readily obtained, used the broader perspectives offered by the Reason of Plato to envisage possibilities. Pumps were already available, but current sources of power were not strong or steady enough for the new needs. The solution involved taking a closed vessel, like Hero's Fountain, supplying pressure, not by blowing in air but by heating water to create steam, and then substituting for the amusing play of water the hissing clank of a powerful piston.

Steadily the spread of theoretical ideals led to the improvement of designs and the invention of new technologies. Closer fitting valves were needed, as were finer tolerances for machined parts, if steam pressure was not to be wasted through gaps and cracks. With Plato's yen for detail and Ulysses' quest for efficiency, new precision devices like Wilkinson's boring machine[12] were deliberately sought and tailor-made to serve both ideals. Carnot (1796-1832), inspired by the steam engine, succeeded in depicting the theoretically ideal heat-machine. Influences began to flow both ways, from practice to theory and again back to practice, with increasing speed and regularity. Then with the creative theoretical power to imagine fresh needs, for which currently imagined devices could be offered as solutions, the steam pumping engine could be transformed, with a device for turning reciprocating motion into rotary motion, into a substitute for wind and water power in mills.

[11]Lewis Mumford, *Technics and Civilization* (New York: Harcourt, Brace, and World, 1963), chap. IV. By permission of Simon and Schuster.
[12]*Ibid.*, pp. 160-1.

After that, by simply putting the engine on a ship or on a platform with wheels, and given the means of imparting rotary motion from the engine to the paddle wheel or to the wheels below, the steam boat and the locomotive could take their places on the waterways and landscapes of Europe and America. The toy of Hero, unlinked to generalized theorizing in pursuit of practical effectiveness, remained a pneumatic toy. Ulysses, however, with a consciousness newly raised by the universal perspective of Plato, used the power implicit in Hero's toy to steam into the modern age.

4.3. **INTELLIGENCE UNITED**

Despite the usefulness of our key distinctions between practical and theoretical intelligence and between technology and science, we must not forget that in the real world these distinguished aspects of reality are not finally separated. They are interactive in various ways, and in the end their interactions are the most important thing about them. Consider the following:

(a) The initial and most inclusive stage is *sheer mentality* itself: the capacity to take account, with favor or disfavor, of what is not physically present. This is the power that organisms with intelligence have to *pursue an idea:* i.e., to act out of interest in a formal possibility, to relate to the actual through an abstraction from the actual.

(b) The next stage is *mentality regulating itself:* the capacity to take account of the proliferation of available possibilities for immediate action, to grade or evaluate them, and to retain repertoires of patterns for action under recognizable circumstances, i.e., to develop methods.

What we see in these two basic stages is the activity of mind as critic of actuality (standing over against actuality with an idea of what could be different) and also the activity of mind as critic and judge of its own criticisms. Practical intelligence is the power of some living organisms to act on the environment with the special leverage of an ideal possibility. Methods of achieving these ideal possibilities are themselves formal possibilities. They are possibilities to be activated by the organism when appropriate for the sake of other possibilities desired or demanded by life. Methods may be simple or complex. They may be highly effective or barely effective or even partly self-stultifying. Once they are achieved they are likely to be retained if they have some effectiveness and if nothing better is at hand. Rote methodologies are time- and effort-saving devices. After the moment of discovery or creation, there comes a period of consolidation and repetition, followed—unless something prevents it—by a long slide into gradual decay of information.

(c) The third stage is *mentality breaking loose from its own achievements:* the capacity to leap free even of its own practical successes and to play with ideas for their own sake. The satisfactions of this play can be a direct enhancement of life. The unbridled anarchy of this play if unregulated, however, may bring the pain of mental chaos. At a minimum it loses the aesthetic satisfactions of unification. At worst it may amount to insanity.

(d) The fourth stage is *mentality disciplining its own mental play:* the capacity to criticize and order its own free creations by method in thinking. This, generally speaking, is logic. Some tend to confuse logic with the whole of reason, but this is too narrow. Logic is to theoretical reason what practical reason is to sheer mentality. It is the self-criticism, ordering, and evaluating of the widest domain of pure formal possibilities, abstract from every actual circumstance and need. Like any method, a logic is itself a complex of formal possibilities, a repertoire of mental procedures to be applied under circumstances that are mentally judged to be appropriate. Like any method, the methods of some specific logic can be more or less effective. Like any method, the methods of a logic tend to be conserved and perpetuated by successful practitioners. And like any method, the limits of a logic will always be subject to the gadfly of free theoretical reason, which is not exhausted by its methods but uses them to refine and reform themselves in the never-ending process of self-transcendence.

The unitive pattern should be apparent by now. Mentality as an agency of life is variously differentiated, but always remains mentality. On the practical level, it is both the anarchical appetition after envisaged possibilities *and*, as practical intelligence, the self-critical generator and defender of methods. It is thus what yearns for improvements over whatever is actual *and* what tenaciously conserves achieved methods, even to the point of obscurantism. On the theoretical level, it is both the free speculative play with formal abstractions *and*, as logic, its own hawk-eyed referee. It is its own most stubborn impediment *and* its own most impatient critic, all in the endless quest for ever-increasing richness and wholeness of life.

4.4. THE EPISTEMOLOGICAL AUTHORITY OF TECHNO-SCIENTIFIC THINKING

One impressive result of the bonding of practical and theoretical intelligence during the last centuries is the advance of reliable understanding. For centuries, theoretical reason yearned for understanding of the universe in all its scope and all its detail. For centuries a vast variety of theories proliferated, with varying degrees of plausibility. Logical argument was used to oppose and defend alternative points of view. Is the whole universe made of one kind of stuff or of many? If one kind, then what is it like? Is it like water,[13] capable of different changes of state from liquid to solid or gas? Or is it like nothing in particular,[14] an undifferentiated stuff bereft of all determinate qualities in order to be capable of supporting them all? If of many

---

[13] John Burnet, *Early Greek Philosophy* (Cleveland, Oh.: World, 1961), pp. 40–50.
[14] *Ibid.*, pp. 50–71.

kinds, do they abound in infinite variety[15] or do they constitute a finite[16] set of basic elements? Are the fundamental elements themselves atomic bits incapable of change,[17] accounting for change merely by geometrical rearrangement? Or is it geometry and mathematics[18] that is itself basic?

The penetration of these questions and the genius with which they were debated and refined for over two thousand years is a proud heritage of unaided theoretical reason. But the implementation of theoretical reason by its association with practical reason has achieved even more.

(a) On the one hand, it has amplified and extended the human experience, and therefore understanding, of nature beyond the wildest dreams of unassisted theoretical reason. In its quest for *minute detail*, theoretical intelligence has been given the means, through instruments such as microscopes, to peer into the nearly infinitely small. In its quest for *global inclusiveness*, theoretical intelligence has been supplied the means, through telescopes, spectrometers, space probes, computers, etc., to expand its gaze to the almost infinitely large. Other instruments, such as Geiger counters, X-ray machines, radios, etc., work with processes unguessed at by the simply speculating mind, no matter how brilliant its native gifts. We are in a qualitatively different cognitive position regarding nature than was possible at any earlier time in the history of the world.

(b) On the other hand, techno-scientific thinking has harnessed these new cognitive powers to the quest of life for survival and for growth. In literal quantitative terms, on the basic biological level, theory-based technologies of sanitation and food production have allowed the unprecedented multiplication of human biomass on this planet. "Bigger," of course, is not necessarily "better," especially when it comes to population (6.3.d); but the urge to "live well and live better" has for the fortunate been satisfied by utilities, comforts, and opportunities unparalleled in history. The primordial human yen to fly, for example, has been at last achieved. The jaded public, living at the close of the century in which this triumph of techno-scientific thinking was first celebrated, may look on flight as just a faster mode of mass transportation to be endured,[19] but the magic of what the new capacity to fly means for the multi-dimensional enlargement of human experience has not been entirely lost.[20]

What does this mean for the epistemological authority of the techno-scientific way of thinking that has brought us so much? Has the human race, by the blending of theoretical and practical reason, at last paved a single great highway to assured truth?

[15]*Ibid.*, pp. 197–250.

[16]*Ibid.*, pp. 251–75.

[17]*Ibid.*, pp. 330–49.

[18]*Ibid.*, pp. 169–96.

[19]Theodore Roszak, *Where the Wasteland Ends: Politics and Transcendence in Postindustrial Society* (Garden City, N.Y.: Doubleday, 1972, Anchor Books edition, 1973), Chap. 10.

[20]Frederick Ferré, "Urge to Fly," *AOPA Pilot*, 27, no. 9, September, 1984, pp. 107–8.

The answer, if thoughtful, cannot be too simple. On the one hand, the enormous achievements of technology-based science and science-based technology can hardly be overstated. They can in no reasonable way be "dismissed" or "gotten around,"[21] as some might prefer, for one motive or another. The implementation of scientific theory in technology has made it impossible to consider it optional, "like any mythology."[22] The vast extensions of human understanding and manipulation made possible by modern techno-scientific thought can legitimately claim too much authority in sheer detail, coherence, and applicability to be given such short shrift.

"The basis of all authority," as Whitehead wrote, "is the supremacy of fact over thought."[23] This does not mean that thought may not be an important ingredient in what the facts are. But in the end, epistemological authority must be grounded in what we find given, primitive to our reflections or elucidations. When (ideally) everything combines—when our theories issue everywhere in successful methods and their associated artifacts, and when our practices are everywhere coherently and adequately interpreted by our theories—that would be the supreme authority for intelligence. Intelligence finally is one.

Thus the supreme verification of the speculative flight is that it issues in the establishment of a practical technique for well-attested ends, and that the speculative system maintains itself as the elucidation of that technique. In this way there is the progress from thought to practice, and regress from practice to the same thought. This interplay of thought and practice is the supreme authority. It is the test by which the charlatanism of speculation is restrained.[24]

If we apply this test to techno-scientific thinking, it is presently unrivaled. Modern industrial society embodies the practical techniques that have been made possible by theories which, in turn, interpret their success. All our characteristically modern institutions manifest a similar blend of ideas and practices. This proven capacity of techno-scientific thought to create modern civilization is of profound epistemological importance. "In human history, a practical technique embodies itself in established institutions—professional associations, scientific associations, business associations, universities, churches, governments. Thus the study of the ideas which underlie the sociological structure is an appeal to the supreme authority."[25]

On the other hand, techno-scientific thinking is not beyond criticism. One important epistemological principle that is easily lost sight of in the euphoria generated by the practical successes of modern technology is that *practical success is*

---

[21]Theodore Roszak, *The Making of a Counter Culture*, (New York, Doubleday, 1969), p. 215.

[22]*Ibid.*, p. 215.

[23]Whitehead, *The Function of Reason*, p. 80.

[24]*Ibid.*, pp. 80–81.

[25]*Ibid.*, p. 81.

no guarantee of truth (3.2.f and 3.4.d). Successful actions and effective methods may spring from premises that turn out to be mistaken. Logically speaking, the entire interconnecting and mutually supporting structure of modern industrial civilization is no *proof* that scientific theories about the world are true. Psychologically it is hard, living cocooned in our technospheres, to pay much heed to this logical reminder; but epistemologically it is necessary to insist on it firmly. This does not mean that techno-scientific thought is *false*, of course. The *absence of proof of truth* is not the same as the *proof of falsity*. The important point to remember is that technological successes do not put current techno-scientific thinking beyond criticism—if theoretical criticism should be needed.

In one sense, criticism is always needed, in principle. The rhythms of rationality (4.3) call for the gadfly sting of theoretical intelligence on every methodology, however successful. This is why Whitehead refuses to call even what he identifies as the "supreme" authority for thought a "final" authority.

Even this supreme authority fails to be final, and this for two reasons. In the first place the evidence is confused, ambiguous, and contradictory. In the second place, if at any period of human history it had been accepted as final, all progress would have been stopped. . . . Nor can we accept the present age as our final standard. We can live, and we can live well. But we feel the urge of the trend upwards: we still look toward the better life.[26]

In what ways are the currently dominant theories of techno-science open to debate? The second half of this book will be devoted, in one way or another, to answering this question. In general, however, two broad areas are especially debatable.

First, the quest of theoretical intelligence for *coherence*—for a unity within which there are no contradictions, no conceptual "levels" with different concepts untranslatable into one another, no theoretical fragments—is today far from satisfied (1.4.b). Even in individual sciences, notably in physics as well as in biology, sociology, psychology, etc., there is no unifying whole. Much less is there a grasp available, or even vaguely indicated, that would unify all understanding.

Second, the demand that theory be *adequate* to all the data—without forced reductions, denials, or explaining away—may be in need of the gadfly sting (1.4.c). Especially in dealing with subtle but important aspects of experience which may suggest the presence of purpose in nature, freedom in nature, or value in nature— issues that were special victims of the historical battle of modern science against pre-modern approaches—failures in adequacy may be suspected. It is a historical contingency, a matter of luck, that the skies of the earth are clear of permanent clouds, like those of Venus, and that the early foundations of techno-scientific theory could thus be planted in astronomy, a domain where simple ideals like frictionless motions (the planets) could establish successful methods of explanation and application. In many ways, it was good luck for the history of modern thought,

26*Ibid.*, p. 81.

since other approaches and subject matters might have been too complicated or too elusive for the establishment of a powerfully implemented theoretical methodology. But methods, once contingently achieved, have a way of perpetuating themselves, sometimes even at the cost of needed improvements. Critics of today's dominant techno-scientific thinking might be tempted to say, with Whitehead:

A few generations ago the clergy, or to speak more accurately, large sections of the clergy were the standing examples of obscurantism. Today their place has been taken by scientists—
By merit raised to that bad eminence.
The obscurantists of any generation are in the main constituted by the greater part of the practitioners of the dominant methodology. Today scientific methods are dominant, and scientists are the obscurantists.[27]

Whitehead may have overstated the case for dramatic effect. But he suggests some important questions. What are the fundamental presumptions of modern scientific and technological methods? What general kind of society and what sorts of institutions do they reinforce and interpret? Is there a techno-scientific "inertia," so conservative about characteristic modern methods and presumptions that serious development of "post-modern" alternatives is discouraged?

In the last chapter of this book, these questions and others will again occupy us directly (8.5 and 8.6). First, however, we need to try to unpack from the *institutions of modern existence*, in general, what may be implicit in, and characteristic of, modern patterns of techno-scientific intelligence. We shall begin in the next chapter by sampling some prominent and varying perspectives. Then we need a more fine-grained examination of specific *ethical problems* raised by current implementations of modern intelligence, problems that lead to *religious debates* over the legitimacy of modern technology as such. In the light of all this we can finally return to ask again about the authority of, and the chance of an alternative to, techno-scientific thinking.

CHAPTER 5

Technology and Modern Existence

5.1. THE TECHNOLOGICAL PHENOMENON

The pervasive "technosphere" with which this book began turns out to be the implementation of a kind of mentality that is characteristic of our modern age. It rests on the unique blend of techno-scientific intelligence that both *creates* the modern technological phenomenon, with its theory-based practical knowhow, and *interprets* it, with its preferred categories of explanation.

This technological phenomenon requires careful philosophical critique on both its practical and theoretical aspects. The rest of this book will indicate broad lines along which such critiques might proceed and will lay out some of the key issues for debate.

The present chapter will survey four different global assessments—two stressing the bright and two the somber side—of the modern technological phenomenon as a whole. Chapter 6, in contrast, will deal with norms of ethical and other kinds of assessment and with questions posed by particular major technologies. Chapter 7 will examine some central issues of religion as these relate to technological civilization. Chapter 8 will examine metaphysical questions raised by the technological phenomenon—and at the end will reopen the theoretical question (4.4) of the "finality" of the modern techno-scientific method of thought.

5.2. BRIGHT VISIONS (1): KARL MARX

Although Marx (1818–83) did not develop a detailed philosophy of technology or write compactly about the technological phenomenon, his ideas on our modern world and the significance of implemented practical intelligence are important in any survey of seminal thought on our topic. Besides their intrinsic philosophical interest, they have indirectly, through the later development of Marxism and its place in world politics, played a major role in shaping the views and circumstances of millions of persons.

a. The Primacy of the Practical

A strong theme in Marx's thought, to put it in the terms we have now developed, is the primacy in human life and history of the Reason of Ulysses. The whole human race, Marx argued, can hardly be characterized by reference to such a recent appearance as theoretical intelligence. Our species, on the other hand, is qualitatively special in the extent of our creation and use of implements and techniques to meet our needs. We are preeminently the tool-using animal.

In this approach, Marx sharply broke with the dominant philosophy of his day, the idealism of G.W.F. Hegel (1770-1831), which advanced the primacy of Spirit not only within human history but also for the universe as a whole. The proper approach, Marx believed, was the complete inverse of Hegel's. The fundamental, he asserted, is the material, not the spiritual. For the real driving forces in history, look to the material challenges and conditions of life, particularly to the universal needs of living things to provide for their own subsistence. These needs and our methods of meeting them are in human terms the economic realities. It is these realities that stimulate the responses of Ulysses. They are what actually rule the world, however much the fact may be disguised by the glib-tongued Plato.

The abstractions of theoretical intelligence, according to Marx, are themselves products of the underlying economic realities. Such abstractions—philosophical systems, cultural styles, political theories, religious beliefs—have no force of their own. Rather, they are "ideological" in the sense that they reflect, directly or indirectly, the material interests of those who think them. Parliamentary government, for instance, is for Marx a by-product of the revolt of capitalism against the feudalistic economic order.[1] So, likewise, is the rise of Protestantism.[2]

b. Technology in Historical Change

The tools and implements that go into the economic methods of any historic system make up a large part of what Marx called the "forces of production." In addition there are the material circumstances that practical intelligence finds at hand, such as available raw materials or geographical features (rivers, fertile fields, and the like), that should not be classed as technology itself. These forces of production are always in interaction with the "relations of production," which are the social

---

[1] Karl Marx, *A Contribution to the Critique of Political Economy* in *Selected Works* by Karl Marx and Friedrich Engels (New York: International, 1968), "Preface," pp. 181-85.

[2] *Ibid.*, pp. 181-85. This theme has been much amplified in distinguished treatments by Max Weber's *Die protestantische Ethik und der Geist des Kapitalismus*, and by Richard H. Tawney's *Religion and the Rise of Capitalism* (New York: Harcourt, Brace and Company, 1926).

arrangements between persons engaged in economic activity. These arrangements need not be, but historically usually are, exploitative relations, as between master and slave, lord and serf, owner and employee.

When the relations of production provide an incentive for the innovative activity of practical intelligence, Marx believed, then (and only then) new technologies arise in human history. The relations of production for most of human history (prior to the modern capitalist era) have tended toward technological stability, but occasionally major changes have occurred. When they do occur, revolutionary changes are in store. The relations of production that stimulate the new forces of production represented by technology are threatened by their own offspring. Technology, once born, outgrows its social cradle and shatters the relations of production that called it forth.

This is because, for Marx, the new forces of production carry within them the potential to expose the contradictions, the internal weaknesses, of the earlier relations of production. Feudalism was built on the fealty between serf and lord. The serf was not a slave. The master could not dispose of serfs like property. The master had duties as well as privileges. But the serf was bound to the support of the master and to the tilling of the feudal land. Within this equilibrium the economic system could persist over long periods. But this equilibrium could not endure the economic need for the special skills—the crafts of the glassmakers, the swordsmiths, the jewelers—that were wanted by the lords. Those crafts required some persons to be free from the constant tilling of the land and led to the growth of towns in which guilds of free craftsmen could work their arts and in which trade could flourish. The internal economic incentives of feudalism, represented by the self-interest of the feudal rulers, were self-contradictory and self-destructive, since they called for technological implementation in towns that grew to be cities of free agents—agents free from bondage to the land, some free to trade and become rich, others free to sell their labor to the highest bidder. Thus feudalism led paradoxically to its own destruction by encouraging the rise of cities in which, eventually, the bourgeoisie would form capitalism.

c. Modern Technology

Capitalism, and the modern world of constantly progressing technology, was brought into being, in dialectical revolt against the contradictions of the feudal age, by the invention of machines and factories, built by the wealthy who thereby became still wealthier through monopolizing the most effective forces of production and (in a new, uniquely capitalist relation of production) buying the labor of unattached workers at a shameful fragment of its worth.

Just as capitalism, for Marx, contains its own internal growth incentives for constantly new inventions and implements, incentives for "constant revolutionizing of production, uninterrupted disturbance of all social conditions, everlasting uncertainty and agitation,"[3] so also it contains its own internal contradictions, the seeds of its own demise, in which technology also plays a key role. The needs of the capitalist rulers lead to ever-increasing concentrations of workers, as industrial technologies in the service of profits grow larger and larger, with new markets to feed—and feed upon. At the same time, the needs of capitalism also have led to widespread literacy, made *possible* by the technology of printing and made economically *necessary* by the factory environment in which workers need a modicum of education to function at a profitable level of efficiency. The two technological forces in combination are an explosive mixture. As the exploited workers are forced together in greater numbers by technologies of the industrial system, they also are made aware, through technologies of educational and mass communication systems, of their exploitation, of the injustice of their circumstances, and of their raw power to revolt against the masters.

d. Technology and the Future

Marx believed that the revolt of the exploited proletariat would bring a new economic order in which, eventually, there would be no contradictions between the forces and the relations of production. This would require technology at its best and most productive, to assure the post-capitalist world of conditions of material plenty such that no one would lack the goods of life. The goal would be a society in which it would be possible to ask each member of society to work "according to his capacity," while providing for each "according to his needs."[4]

In this way, the modern technological phenomenon acts not only as a goad to profoundly needed changes in society, an intensifying stimulus to revolution against the contradictions within capitalism, but also as a lure to a new equilibrium of productive forces and social arrangements that will fulfill all human hopes. The attitude of Marx toward technology is finally hopeful. Technology does not tell the whole human story, of course, since social arrangements and natural resources are no less important factors in the economic outworking of things, but it is an indispensable and positive ingredient in the dynamics of the material dialectic of history.

5.3. BRIGHT VISIONS (2): BUCKMINSTER FULLER

R. Buckminster Fuller (1895-1983) was himself a striking phenomenon of the twentieth century. He was an engineer, an inventor, a philosopher, a prophet, and an indefatigable public speaker even in his advanced years. Fuller's best known invention was the geodesic dome, which achieved, through its geometric structure,

[3]Marx and Engels, *Selected Works*, p. 38.
[4]*Ibid.*, pp. 333-49.

remarkable strength per unit weight. Fuller's light and airy dome over the United States Pavilion at Expo '67 in Montreal was one of the notable sights of that world's fair.

Among other things Fuller designed his Dymaxion house and three-wheeled Dymaxion automobile, all with keen attention to mathematical relationships, function, and efficiency. It was Fuller's constant contention that by taking the right kind of thought one could do "more with less." His 90 horsepower automobile, for example, was capable of driving at 120 miles per hour.

He also reflected extensively on his inventions and designs, pushing his thought to cosmic dimensions. In the process he sometimes modified normal language patterns to suit his ideas and wrote poetry to help communicate to his often befuddled public. The poetry arose from his attempt to provide clarity by inserting breaks and pauses in the otherwise torrential flow of his language. He referred to it as "mental mouthfuls of ventilated prose"[5] in the preface of one of his most wide-ranging and fascinating books. Most of the following discussion of his unique vision will be documented in his highly personal linguistic form.

a. The Cosmic Primacy of the Theoretical

For Fuller the distinguishing mark of the human is the Reason of Plato. The rest of the universe is "spontaneous" but the human race is increasingly reflective and aware. "The history of man seems to demonstrate the emergence of his progressively conscious participation in theretofore spontaneous universal evolution. Man seems unique in this progressive degree of conscious participation in evolution."[6] Such reflective power is for Fuller immensely practical as well. It is the capacity that not only gives us our special status but also helps those who trust it to survive.

One of Fuller's vivid images is of the instrument pilot (or the submariner also operating on instruments) who succeeds because scientifically designed instruments are trusted rather than bodily sensory signals. In the title poem of his book, Fuller meditates on the outbreak of fighting at the start of the Second World War. An air and sea battle has occurred near Oslo.

> *I think of such of the aviators and sailormen as*
> *are in command of their faculties*
> *on both sides at this moment.*
> *Though you have been out in*
> *a froth-spitting squall*
> *on Long Island Sound or*
> *in an ocean liner on a burgeoning sea*
> *you have but a childlike hint of*

[5]R. Buckminster Fuller, *No More Secondhand God and Other Writings* (Garden City, N.Y.: Doubleday, 1963, Anchor Books edition, 1971), p. x. With the permission of the Buckminster Fuller Institute, Los Angeles, CA.

[6]*Ibid.*, p. viii.

what a nineteen-year-old's reaction is
to the pitch black shrieking dark out there
in the very cold northern elements
of unloosening spring
off Norway's coast
tonight
15,000 feet up, or
fifty under or
worse,
in the smashing face of it and
here I see God.

I see God in
the instruments and the mechanisms that
work
reliably,
more reliably than the limited sensory departments of
the human mechanism.

And he who is befuddled by self or
by habit,
by what others say,
by fear, by sheer chaos of unbelief in
God
and in God's fundamental orderliness
ticking along on those dials
will perish.
And he who unerringly
interprets those dials
will come through.[7]

Every instrument pilot will testify that one must learn to mistrust the senses and to trust the gauges even when every nerve insists that they must be mistaken. A good instrument instructor will make sure that the trainee pilot experiences such a conflict under safe conditions. This can be easily induced by the instructor's banking the plane steadily for a while, unknown to the trainee, then swifly rolling the aircraft into level flight and immediately asking the trainee to fly by reference to the instrument indications alone. One's organs of balance become quickly adjusted to the simulated gravitational forces in a coordinated turn and, if suddenly made level, they will register to consciousness a false sense of being thrown into a turn in the opposite direction. The instruments will indicate a level flight condition, but one's body screams to tilt the plane and make it "level." One must learn to believe one's mind—fed by the cool eyes that read the instruments—rather than

[7]*Ibid.*, pp. 1–2.

one's body. Otherwise, in real conditions of turbulence and distraction, the flight will doubtless end sooner and sadder than intended.

The increasing instrumentation of the human race by theoretical intelligence is for Fuller what provides our species its special powers and destiny. We now know that the universe is full of phenomena that are undetectable by our naked senses. But theoretical intelligence, implemented with high technology, gives us access to these otherwise unknowable dimensions.

Man is born with an extraordinary inventory of faculties within an extraordinary inventory of universal phenomena. Most of the inventory is invisible, operating either infra or ultra to our sense apprehending. My philosophic working assumption goes on to assume that, despite the meager degree in which we consciously employ our capabilities in response to the meager degree in which we understand the universal phenomena, we were given our faculties to permit and induce our progressively greater apprehension and comprehension of the universal phenomena.[8]

b. The Cosmic Setting of Technology

The phrase "we were given" in the previous quotation is more than a way of speaking for Fuller, since the conceptual depth and orderliness of the universe strongly suggest to him a prior coordination between our theoretical capacities and the universe itself, from which we spring. Science and technology, as Fuller sees them, are at work exploring what has already been anticipated by intellect in nature. Fuller writes about "what seems to me to be an overwhelming confrontation of our experience by a comprehensive intellect magnificently greater than our own or the sum of all human intellects which has everywhere and everywhen anticipatorily conceived of the complex generalized, fundamental principles which all together interact as universe."[9] And if this is so, then the crucial place of modern technology for the universe as a whole becomes clear. "My continuing philosophy is predicated . . . on the assumption that in dynamical counterbalance of the expanding universe of entropically increasing random disorderliness, there must be a universal pattern of omnicontracting, convergent, progressive orderliness and that man is that anti-entropic reordering function of universe."[10]

Human destiny is centrally bound up with the techno-scientific enterprise. Fuller sums up poetically:

> . . . tonight vividly (as tacitly always)
> God is articulating
> through his universally reliable laws.
> Laws pigeonholed by all of us
> under topics starkly "scientific"–

[8]*Ibid.*, p. vii.
[9]*Ibid.*, p. vii.
[10]*Ibid.*, p. vii.

behavior laws graphically maintained in the performance
of impersonal instruments and mechanics
pulsing in super sensorial frequencies
which may serve yellow, black
red, white, or pink
with equal fidelity.
And I see conscious man alone
as mechanically fallible
and progressively less reliable
in personal articulation
of God's ever swifter word,
which was indeed in the beginning.
Only as mind-over-matterist,
as philosopher, scientist,
and informed technician
impersonally and universally preoccupied
is man infallible.[11]

c. The Historic Challenge of Technology

If God is thought of "as a verb"[12] for the active reordering processes of the universe, and human beings as "trans-ceiver mechanisms through which God is broadcasting,"[13] then the characteristic economic and political institutions in modern technological society can be fully understood for the first time.

Engineering is seen by Fuller as the profession at the leading edge of history, since it is the engineer who incorporates ideas into the material and social order. In his poem "Machine Tools," Fuller celebrates the creation of an airplane out of the elements of the "raw countryside"—an anti-entropic process of ordering the scattered elements. But there is a key link in the process, which could threaten the whole cosmic enterprise: namely, the paucity of good mechanical engineers who can produce the machine tools to support the technological edifice.

In converting one hundred tons
of raw broad countryside
into five tons
of scintillating airplane-in-flight,
the machine-tool is specifically
that link
in the industrial chain of events. . . .

[11]*Ibid.*, p. 17.
[12]*Ibid.*, p. 23.
[13]*Ibid.*, p. 26.

Here are prepared
the mechanical surfaces
between which time and energy
are masculated.
And here man and his wisdom
must be the master.
Yet there are few of his members
qualified for such mastery.
And there creaks incisively
today's weakest link.[14]

Fuller's own commitment was not to theory alone, despite his great respect for theory. Just as he wanted more and better engineers, he wanted the results of theory to show in industry and economics.

The individual intellect disciplinedly paces the human individual. The individual disciplinedly paces science. Science disciplinedly paces technology by opening up, both widened and refined, limits of technical, advantage generating, knowledge. Technology paces industry by progressively increasing the range and velocity inventory of technical capabilities. Industry in turn paces economics by continually altering and accelerating the total complex of environment controlling capabilities of man. Economics in turn paces the everyday evolution acceleration of man's affairs.[15]

In political as well as economic institutions, modern scientific technology could decisively "pace" or lead the way to an unprecedented reawakening of democracy, Fuller believed. The forces of anti-democratic repression in the world could be effectively answered, as never before in history, if the technical possibilities of providing direct democratic decision making were only tapped.

Democracy has potential within it
the satisfaction of every individual's need.

But Democracy must be structurally modernized
must be mechanically implemented
to give it a one-individual-to-another
speed and spontaneity of reaction
commensurate with the speed and scope
of broadcast news
now world-wide in seconds.

Through mechanical developments
of the industrial age

[14]*Ibid.*, pp. 38–39.
[15]*Ibid.*, p. ix.

the cumulative production of human events
within the span of a four-year administration
is now the quantitative equivalent
of the events of a four-hundred-year
pre-industrial dynasty.[16]

There is likely to be a penalty imposed upon the human species if we miss this historic chance to meet our destiny by embracing fully the technological future, Fuller warns. We need to honor the engineer, stimulate the economy through disciplined scientific and technological change, and transform our political life into a real democracy that will unlock enough free human energies to rid the world of tyrants. If we fail, then the cosmic process may go on without us. At this "unique threshold moment in history,"[17] it is not clear which way our species is going to go. In the tradition of older prophets, but with a more humorous twist at the end, Fuller speculates that if we fail in our cosmic task, God will not forever be frustrated.

For God may reasonably be
slowly up-winding
that game of shoot-the-works
through the instrumentality of man;
or failed by man,
possibly through
some other animate specie or process
like aurora borealis
cosmic electrolysis.[18]

5.4. SOMBER VISIONS (1): MARTIN HEIDEGGER

The evocative thought of Martin Heidegger (1889–1976) has deeply influenced many philosophers of the twentieth century, particularly those who stress phenomenological description of the experienced structures of human existence (1.4.c). These concerns, sometimes called "existentialist," were already present in Heidegger's major work, *Being and Time* (1927)[19] and were brought fully to bear

---

[16]*Ibid.*, p. 9.

[17]*Ibid.*, p. 95.

[18]*Ibid.*, p. 34.

[19]Martin Heidegger, *Being and Time*, trans. John Macquarrie and Edward Robinson (New York: Harper & Row, 1962). Excerpts totalling approximately 525 words passim from *The Question Concerning Technology and Other Essays* by Martin Heidegger, translated by William Lovitt. Copyright © 1977 in the English translation by Harper & Row Publishers, Inc. Reprinted by permission of Harper & Row Publishers, Inc.

on the theme of technology in Heidegger's thought-provoking essay, "The Question Concerning Technology" (1954).[20]

A pioneering voice, Heidegger was initially interpreted by some as merely negative toward technology, yearning romantically for a bygone age of craft traditions and windmills. There is some basis for this interpretation in Heidegger's essay, as we shall see, but although Heidegger's view of the modern technological phenomenon is somber and filled with warnings, he was not simply "anti-technological." The rich complexity of his approach, especially his fascination with what can be suggested by Greek etymologies and German stems, cannot possibly be captured in a summary. Still, we must do our best since the depth of Heidegger's questioning requires careful attention.

a. The Essence of Technology

The primary "question concerning technology," Heidegger asserts, is "what it is."[21] In a sense that question can be quickly handled, he acknowledges, by simply giving a definition blending two widely recognized aspects of technology: first, that it is an *end-seeking* human activity and, second, that it is the use of *equipment, tools, machines,* and the like, to achieve those ends. The elements of this definition are in keeping with our own (2.7), since "end-seeking" is another way of referring to the "practical" side of human affairs, and "equipment", etc., is covered by the term "implementations" in our definition of technologies as practical implementations of intelligence.

Heidegger acknowledges the "correctness" of such an "instrumental and anthropological"[22] definition of technology. But here he makes a useful distinction between what is "correct" and what is "true." The distinction is in some ways similar to Whitehead's warning against "the fallacy of misplaced concreteness,"[23] that is, the tendency to confuse a significant part of a thing with the whole concrete reality from which the part has been abstracted. As Heidegger puts it:

The correct always fixes upon something pertinent in whatever is under consideration. However, in order to be correct, this fixing by no means needs to uncover the thing in question in its essence. Only at the point where such an uncovering happens does the true come to pass. For that reason the merely correct is not yet the true.[24]

Nevertheless, the "correct" definition shows its correctness by covering the whole technological domain, both old and new. Craft technologies are means to practical

---

[20]Martin Heidegger, *The Question Concerning Technology and Other Essays*, trans. William Lovitt (New York: Harper & Row, 1977).

[21]*Ibid.*, p. 4.

[22]*Ibid.*, p. 5.

[23]A.N. Whitehead, *Science and the Modern World* (New York: The Free Press, 1929, paperback edition, 1967), p. 51.

[24]Heidegger, *The Question Concerning Technology*, p. 6.

ends, but so are science-led technologies like power plants and jet aircraft.[25] Thus if we think entirely in terms of the anthropological-instrumental definition, we shall attempt to relate ourselves to modern technology as a mere means, something to be manipulated for our practical ends and kept firmly under our human mastery.

There is something wrong here, however, since in other respects modern technology is "something completely different, and therefore new."[26] It is radically different to the extent that the modern technological phenomenon *challenges* nature in a way that the older technologies never did. High technology demands the extraction of energy from nature for storage and manipulation at will. Heidegger puts the contrast with the old as follows, "But does this not hold true for the old windmill as well? No. Its sails do indeed turn in the wind; they are left entirely to the wind's blowing. But the windmill does not unlock energy from the air currents in order to store it."[27]

This qualitatively different character between modern and traditional technologies shows that the essence of technology has not really been uncovered in the "correct" definition. Something much more fundamental is at stake. Heidegger looks for help to the ancient Greek understanding of *technē*, and finds that all *technē* was primally a "bringing-forth," and not merely the causal "bringing-forth" of instrumental crafts and skills but also the creative "bringing-forth" of the fine arts as well. Plato associated *technē* with *epistēmē* as a kind of knowing, a being expert and at home in some area, an understanding, an opening up or revealing. Aristotle refined this concept to refer to the "bringing-forth" that occurs when such revealing does not occur naturally, as in a living organism.

Whoever builds a house or a ship or forges a sacrificial chalice reveals what is to be brought forth. . . . Thus what is decisive in *technē* does not lie at all in making and manipulating nor in the using of means, but rather in the aforementioned revealing. It is as revealing, and not as manufacturing, that *technē* is a bringing-forth.[28]

b. The Technological *a Priori*

Certain key questions now need to be asked. Is modern technology, though a form of *technē*, a completely new phenomenon, compared to the old craft technologies with which it shares its name? If so, is it also a "revealing"? Heidegger answers that modern technology is definitely new, not only in its demands on nature but also, as we have seen (3.4, 4.2, and 4.3), in its intimate relationship with modern science. On the one hand, modern technology is "based on modern physics as an exact

[25]*Ibid.*, p. 6.
[26]*Ibid.*, p. 5.
[27]*Ibid.*, p. 14.
[28]*Ibid.*, p. 13.

science,"[29] but on the other hand, "modern physics, as experimental, is dependent upon technical apparatus and upon progress in the building of apparatus."[30]

Still, a deeper question demands answering: "Of what essence is modern technology that it happens to think of putting exact science to use?"[31] Before there can be an attempt to manipulate and control nature by the exact laws discovered in science, there must first be the inclination to manipulate and control nature with ever greater efficiency. Here is what could be called the technological *a priori*, which is not itself a machine or anything overtly technological but is the "machine way of thinking" that allows nature to be approached as something to be mechanized.

This characteristically modern way of thinking and experiencing, Heidegger holds, rather than any overt techniques or artifacts, is the essence of modern technology; and this *a priori* framework of manipulation, control, and "setting-in-order" is what modern technology reveals. What Buckminster Fuller praised (5.3) as the essence of human destiny, the "anti-entropic ordering function" of implemented intelligence, Heidegger also identifies as essential. But for Heidegger this essence belongs uniquely—and disturbingly—to our era. In earlier days, the earth was cultivated, cared for, and maintained; now the earth beneath what was the peasant's field is "challenged" for its mineral deposits, which are stored and ordered, mined for our use. Suppose the mine is coal. In effect, modern technology "challenges" and stockpiles the sun's former warmth, then orders it to deliver the steam that keeps our machinery running. Or suppose that the field is left in agricultural use. Modern technology no longer allows the seed to be put "in the keeping of the forces of growth"[32] as did farmers from earliest days until now; instead agriculture is "the mechanized food industry."[33] Everywhere, Heidegger concludes, the essence of technology appears; the "Enframing" of the world, as a manipulable "standing reserve" for being ordered and regulated, takes place.

c. The Danger of Technology

One usually thinks about the risks and threats posed by modern technology in terms of the possibility of nuclear catastrophes or the like, but for Heidegger the main danger lies much deeper. It lies, paradoxically, in the fact that the essence of modern technology, as a way of revealing how things can be, is the revealing of some of the truth. "Being" is, at least at one level, revealed as amenable to manipulation and control. Thus modern technology cannot be dismissed and must not be underestimated, nor is it amenable to the kinds of controls we usually recommend. Too often, when we start to wonder whether modern technology is out of our con-

[29]*Ibid.*, p. 14.
[30]*Ibid.*, p. 14.
[31]*Ibid.*, p. 14
[32]*Ibid.*, p. 15.
[33]*Ibid.*, p. 15.

trol, we exhort ourselves to " 'get' technology 'spiritually in hand.' "[34] We want to "master" it. But if the technological *a priori* is the *will to mastery* itself, then our firmest determinations will merely pour fuel on the all-consuming flames of the modern technological phenomenon. The more we will to master it, the more it masters us through the technological quality of our act of willing.

Why, though, should the essence of modern technology be resisted? It is, after all, a "destining of revealing"[35] that is given to our time as a mode of experiencing and relating to our world. Should we not simply embrace our destiny and live whole-heartedly in the technological world?

Heidegger's negative answers, through couched in terms which sometimes appear to be either aesthetic or moral, are always at bottom rooted in ontology. On the one hand, he appeals to the intuition of loss or even desecration that comes from taking everything in nature, even the mighty Rhine River, as mere "standing-reserve" for our command. He portrays a hydroelectric plant set into the Rhine. Everything is orderly, the machinery whirs and electricity for our use is produced from the turbines pushed by the current. The Rhine is now something for human disposal. It is not the same river that was spanned by the old wooden bridges. It has been changed by being dammed from a free-running flood used respectfully by boaters into "a water power supplier."[36] Heidegger is clearly affronted by the change, calling it "monstrous." "In order that we may even remotely consider the monstrousness that reigns here, let us ponder for a moment the contrast that speaks out of the two titles, 'The Rhine' as dammed up into the *power* works, and 'The Rhine' as uttered out of the *art* work, in Hölderlin's hymn by that name."[37] Even if technological consciousness appears still to appreciate the river as landscape, it will only be as transformed into an aesthetic commodity available for purchase by tourists.

On the other hand, Heidegger worries that the tendency to approach everything as "standing-reserve" is tending to reduce human beings, like all else, to "human resources."[38] This tendency, while serious, may have a natural limit, however, and we shall return to this later.

Ontologically, Heidegger observes that whatever sorts of objects are at stake, the very status of *object* itself—something standing firmly over against us, just being what it is—is lost by the reduction of all things to mere "standing reserve." Everything is ordered about. Nothing retains the integrity of being something with its own independence of the technological *a priori*. Even large shiny items we can ride in cease to be objects in this sense.

Yet an airliner that stands on the runway is surely an object. Certainly. We can represent the machine so. But then it conceals itself as to what and how it is.

[34]*Ibid.*, p. 5.
[35]*Ibid.*, p. 25.
[36]*Ibid.*, p. 16.
[37]*Ibid.*, p. 16.
[38]*Ibid.*, p. 18.

Revealed, it stands on the taxi strip only as standing-reserve, inasmuch as it is ordered to ensure the possibility of transportation. For this it must be in its whole structure and in every one of its constituent parts, on call for duty, i.e., ready for takeoff.[39]

Finally, and worst of all, Heidegger contemplates the day when the essence of modern technology, as a form of revealing that reduces human experience of everything that is (including human beings themselves) to "nothing but" instrumenality and resources, "drives out every other possibility of revealing."[40] Art, people, even God, would then be taken in this modality as "nothing but" elements in causal chains.[41] And if this process of revealing *as ordering* goes too far, it will have crushed out the possibility of other modes of revealing that may offer deeper reality. "The rule of Enframing threatens man with the possibility that it could be denied to him to enter into a more original revealing and hence to experience the call of a more primal truth."[42]

d. The Grounds for Hope

Heidegger locates the basis for "saving power," despite these profound dangers from the essence of modern technology, in the relationship between the human and the larger-than-human nature of Being. Human existence is unique in the universe for its role in the self-revelation of Being. Even when human nature is threatened by the technological *a priori* with being reduced to standing-reserve ("human resources," "man power," etc.), there is no fear in Heidegger that this fate—despite our coming right to "the very brink"[43]—will ever finally obliterate what is distinctively human. Totter at the brink we may, but "precisely because man is challenged more originally than are the energies of nature, i.e., into the process of ordering, he never is transformed into mere standing-reserve. Since man drives technology forward, he takes part in ordering as a way of revealing."[44]

Still, technology in its essence is not our doing alone. That is the deepest significance of the *a priori* character of this essence. Human beings *find* the technological way of thinking, they do not "make" it. For humans to be in a position even to contemplate "making" such a possibility, it would have to exist already as a possibility for thinking. This is why Heidegger insists that "modern technology as an ordering revealing is, then, no merely human doing."[45] It "sets upon" human beings as a historical destining that is not at our voluntary disposal. Both sides of this process—the essential human role and the essential trans-human context—must be

[39]*Ibid.*, p. 17.
[40]*Ibid.*, p. 27.
[41]*Ibid.*, p. 26.
[42]*Ibid.*, p. 28.
[43]*Ibid.*, p. 27.
[44]*Ibid.*, p. 18.
[45]*Ibid.*, p. 19.

stressed. "Does this revealing happen somewhere beyond all human doing? No. But neither does it happen exclusively *in* man, or decisively *through* man."[46] Likewise, the human condition, though "swayed" by its destiny, is never completely determined by it, since "that destining is never a fate that compels."[47]

Heidegger's limited offer of hope, then, is grounded both in the special role and freedom of the human, and in the mysterious historical destining that is prior to every human response. If human beings can only somehow allow the present technological essence to reveal with enough clarity what it is, there is hope. Heidegger denies that technology is the "fate" of our age,

where "fate" means the inevitableness of an unalterable course. But when we consider the essence of technology, then we experience Enframing as a destining of revealing. In this way we are already sojourning within the open space of destining, a destining that in no way confines us to a stultified compulsion to push on blindly with technology or, what comes to the same thing, to rebel helplessly against it and curse it as the work of the devil. Quite to the contrary, when we once open ourselves expressly to the *essence* of technology, we find ourselves unexpectedly taken into a freeing claim.[48]

The more we look squarely at the danger, then, the more the "saving power" is permitted to "flash" and the more primal truths *may* come clear to the human who is willing to "renounce human self-will"[49] (which is, after all, only another disguise for the technological *a priori* that tends to block all alternative revealings of truth) and in stillness find insight into that which is.[50]

There are no guarantees. The destinings of things cannot be "engineered" or "calculated." The hope for that sort of solution is exactly part of our problem. But neither need we abandon our machines for a primitive pre-industrial life in the Black Forest. That form of romantic anti-technological protest is too superficial for Heidegger's larger insights. No doubt the machines and techniques of an era not dominated by the essence of modern technology will be significantly different from those we now live with. To fill that out in detail would be an interesting speculation. What is important from a Heideggerian perspective, however, is that we ready ourselves to become free from the spiritual "machine" within us that stifles at the centers of human existence. If these crucial issues are squarely faced, the poet Hölderlin offers the best comfort Heidegger can offer:

> *But where danger is, grows*
> *The saving power also.*[51]

[46]*Ibid.*, p. 24.

[47]*Ibid.*, p. 25.

[48]*Ibid.*, pp. 25–26.

[49]"The Turning," *Ibid.*, p. 47.

[50]"The Turning," *Ibid.*, p. 49.

[51]Johann Christian Friedrich Hölderlin, quoted in "The Question Concerning Technology," *Ibid.*, p. 34, and in "The Turning," *Ibid.*, p. 42.

5.5. SOMBER VISIONS (2): HERBERT MARCUSE

As one of the most influential voices in the twentieth century counter-cultural[52] movements of Europe and America, Herbert Marcuse (1898–1979) combined in his own critical perspective several elements from the visions of Marx and Fuller and Heidegger. Like Marx, Marcuse believed that industrial capitalism is radically exploitative of workers; like Fuller, he believed that technological intelligence is capable of ordering a world without poverty; and like Heidegger, he was convinced that the technological *a priori* rules contemporary consciousness and thereby dominates every aspect of political and social life today. In contrast to Marx, Marcuse did not believe that the proletariat, despite their oppression, are likely to rise unaided in revolt against their more sophisticated technocratic masters. In contrast to Fuller, Marcuse did not rejoice in technological progress as in itself the manifestation of human destiny, though in his view only technologically embodied solutions would satisfy the needs of the future. In contrast to Heidegger, Marcuse believed in political critique and activity against the flattening effects of "the happy consciousness."[53]

a. Technology and the Loss of Transcendence

The central problem posed by the modern technological phenomenon, for Marcuse, is its radically engulfing nature. It is, in the root sense of the term, "totalitarian."

By virtue of the way it has organized its technological base, contemporary industrial society tends to be totalitarian. For "totalitarian" is not only a terroristic political coordination of society, but also a non-terroristic economic-technical coordination which operates through the manipulation of needs by vested interests.[54]

The control over machinery in society is political power. That is the dominant new fact of modern civilization. The interlocking political, economic, and technical elites hold total control because society, as never before, is a "rational" system. Everything works together to maximize the smooth running of this system and its unprecedented productivity.

The common people continue to contribute the most to this productivity and to reap the least. They are, as Marx showed, systematically robbed by the system. But the development of the technological society has stolen from the common people even more than the surplus value of their production: it has stolen their awareness of being an oppressed and victimized proletariat. That, for Marcuse, is the most

---

[52]Theodore Roszak, *The Making of a Counter Culture: Reflections on the Technocratic Society and its Youthful Opposition* (Garden City, N.Y.: Doubleday, 1969).

[53]Herbert Marcuse, *One-Dimensional Man: Studies in the Ideology of Advanced Industrial Society* (Boston, Mass.: Beacon Press, 1964), p. 76. Copyright © 1964 by Herbert Marcuse. Reprinted by permission of Beacon Press.

[54]*Ibid.*, p. 3.

insidious development. "All liberation depends on the consciousness of servitude."[55] If consciousness itself is distorted by technological rationality, then servitude itself becomes a hopelessly permanent condition.

How does technological society work this new kind of repression? The new social controls are subtle because they operate at the level of human needs themselves. Right down to basic instinctive needs, like sex, the modern technological order has fabricated a society that keeps people "happy" and docile, not even aware that they are being manipulated and controlled at every point. The mass markets provide people with the "freedom" to choose between this and that item, though not to choose to reject the wasteful consumption of products, since without such constant waste the capitalist profits would be threatened. The centralized mass media indoctrinate even as they entertain. False needs are created by advertising—just the "needs" that more consumption can solve, of course—but real needs, that might dangerously lead to liberation from the meaningless round, are suffocated. Marcuse puts it bluntly:

The distinguishing feature of advanced industrial society is its effective suffocation of those needs which demand liberation—liberation also from that which is tolerable and rewarding and comfortable—while it sustains and absolves the destructive power and repressive function of the affluent society. Here, the social controls exact the overwhelming need for the production and consumption of waste; the need for stupefying work where it is no longer a real necessity; the need for modes of relaxation which soothe and prolong this stupefication; the need for maintaining such deceptive liberties as free competition at administered prices, a free press which censors itself, free choice between brands and gadgets.[56]

The root of the new technological repression lies in consciousness itself, the elimination of the chance for the Reason of Plato to function in its gadfly role. When everything present is affirmed, when everyone is "happy," then imagination itself is crippled in its power to take account of the absent, to long for what *is not.* Marcuse realizes that the surly refusal to "go along" with the rational society must appear neurotic,[57] but to this there are two answers. First, this new identification of all classes with the "rational" society as a whole has been artificially manipulated. It is, Marcuse says:

the product of a sophisticated, scientific management and organization. In this process, the "inner" dimension of the mind in which opposition to the status quo can take root is whittled down. The loss of this dimension, in which the power of negative thinking—the critical power of Reason—is at home, is the ideological counterpart to the very material process in which advanced industrial society silences and reconciles the opposition.[58]

[55]*Ibid.*, p. 7.
[56]*Ibid.*, p. 7.
[57]*Ibid.*, p. 9.
[58]*Ibid.*, pp. 10-11.

Second, the "rationality" itself of the technological society is deeply irrational. It is based on programmed waste, environmental heedlessness, and constant preparation for nuclear annihilation. Even the instant sexual gratifications it offers, vicariously through its slick magazines or directly through its sanitized liberal mores, are reductive from the full sensuality of pre-technologized romance.[59] And the individual is helpless within all this insane rationality.

Domination is transfigured into administration. The capitalist bosses and owners are losing their identity as responsible agents; they are assuming the function of bureaucrats in a corporate machine. . . . Hatred and frustration are deprived of their specific target, and the technological veil conceals the reproduction of inequality and enslavement.[60]

b. Toward the Pacification of Existence

If the totalitarianism of modern technological society is to be fought, it will first of all require the re-stimulation of the lost dimension of imagination and negative critique. The gadfly Reason of Plato will need to be released on all domains of life. Then the machine could be recaptured for the fulfillment of the human rather than *vice versa*. After all, Marcuse writes, "The political trend may be reversed; essentially the power of the machine is only the stored-up and projected power of man. To the extent to which the work world is conceived of as a machine and mechanized accordingly, it becomes the *potential* basis of a new freedom for man."[61]

Marcuse does not hold out a firm promise for the realization of such a potential. But it is at least not impossible if the dynamics within techno-scientific thinking are such as to lead to its own self-transcendence. Perhaps technological progress has its own dialectical rhythm. Suppose, with Fuller, that the trend toward "entropic re-ordering" and "doing more with less" is powerfully possible through human theoretical reason. And suppose, with Marx, that internal contradictions sometimes ripen and burst in surprising and revolutionary ways. Then, Marcuse speculates, once techno-scientific rationality has reached its maximum development on the practical side and has no place to go but toward the free play of art and theory, we might anticipate a qualitative revolution.

At the advanced stage of industrial civilization, scientific rationality, translated into political power, appears to be the decisive factor in the development of historical alternatives. The question then arises: does this power tend toward its own negation—that is, toward the promotion of the "art of life"? Within the established societies, the continued application of scientific rationality would have reached a terminal point with the mechanization of all socially necessary but individually repressive labor. . . . But this stage would also be the end and limit of the scientific rationality in its established structure and direction. Further progress would mean

[59]*Ibid.*, Chap. 3: "The Conquest of the Unhappy Consciousness: Repressive Desublimation."
[60]*Ibid.*, p. 32.
[61]*Ibid.*, p. 3.

the *break*, the turn of quantity into quality. It would open the possibility of an essential new human reality—namely, existence in free time on the basis of fulfilled vital needs. Under such conditions, the scientific project itself would be free for trans-utilitarian ends, and free for the "art of living" beyond the necessities and luxuries of domination.[62]

Such a forecast involves risk, since it requires that the present trends of technological society need to continue to their full completion before they can be transcended. It requires both the "can-do" technological optimism of a Fuller and the revolutionary temperament of a Marx. But if it occurs, it will create material conditions that are unprecedented in history and could allow the transformation of the world into a genuine paradise.

Marcuse insists that this cannot possibly be conceived without a base of continued high technological support. "For it is this base which has rendered possible the satisfaction of needs and the reduction of toil—it remains the very base of all forms of human freedom."[63] But with such a basis, the highest values of our species can be translated into technical tasks[64]—and accomplished. At last there could be material satisfactions without mental repression and "free development of needs on the basis of satisfaction"[65] that would bring about a new respect for individual persons and a new tenderness toward nature. This is what Marcuse called "pacified existence,"[66] the "tabooed and ridiculed *end* of technology, the repressed final cause behind the scientific enterprise."[67]

Can this be realistically hoped for? Marcuse will not give much comfort to those who seek optimistic assurances. We are, it seems, at the beginning of the end of a period in history. "Nothing indicates," he warns, however, "that it will be a good end."[68] The power of current repressive society is too strong to give us much confidence in its demise. The vision of pacified existence is only a chance. If the best thinking for our situation is negative thinking, then let the gadfly sting on, "loyal to those who, without hope, have given and give their life to the Great Refusal."[69]

Summary

This chapter has surveyed a wide range, from the Ardent Embrace of technological society to the Great Refusal. In the process we have sampled important claims and counter-claims. There are many more. Our aim has not been to exhaust or to argue

[62]*Ibid.*, pp. 230–31.
[63]*Ibid.*, p. 231.
[64]*Ibid.*, p. 232.
[65]*Ibid.*, p. 234.
[66]*Ibid.*, p. 235.
[67]*Ibid.*, p. 235.
[68]*Ibid.*, p. 257.
[69]*Ibid.*, p. 257.

with these visions but to share them, momentarily at least, and thus to see from different vantage points the many-faceted technological phenomenon.

Without alternative visions, our minds tend to be locked, without our knowing it, within narrower outlooks based on more limited experience. The assessment of a great issue needs to begin with a recognition of the spread of choices that can and must be made. Then, when our minds have been opened to the breadth of possibilities, we are better prepared for the careful process of critique.

CHAPTER 6

Ethics, Assessment, and Technology

6.1. BASIC ETHICAL THEORY

Before proposing any ethical critique of modern technologies or raising specific questions for ethical discussion, we need to review our available resources and pause to remind ourselves of what such a critique can and cannot hope to accomplish.

Ethical theory is a well-developed field within axiology (1.3.b and 1.8). We can reasonably hope to go well beyond the level of much preference-touting that is often loosely taken for ethical discussion. We should be able to clarify comprehensive issues in a framework that lends itself to critical examination and principled agreement or disagreement.

Ethical theory is not, however, designed to remove disagreements altogether. It may illuminate vital questions for responsible choice, but it will not finally by itself single out a uniquely ethical policy for action. It can push our preferences back to our fundamental convictions about what is real and worthwhile, and thus require us (insofar as we are rational) to acknowledge the implications of our choices, but it is not designed to guarantee moral behavior or even to guarantee unanimity on exactly what moral behavior requires.

In this chapter we shall need to lay out a basic theoretical approach[1] to the principal ethical standards we can apply, contrast this approach with other types of technology assessment now in frequent use, and then sample some of the many

[1] Especially helpful works in the field, to which the present discussion is especially indebted, include William Frankena, *Ethics*, 2nd ed. (Englewood Cliffs, N.J.: Prentice-Hall, 1971); Nicholas Rescher, *Distributive Justice* (Indianapolis, Ind.: Bobbs-Merrill, 1966); and John Rawls, *A Theory of Justice* (Cambridge, Mass.: Harvard University Press, 1971).

ethical questions that clamor for attention in our technological modern world. Every one of these topics could warrant a shelf of books (and some have already generated a large literature), but here they must be taken as exemplary and perhaps as direction pointers.

Let us begin with a survey of the two basic aspects that require attention within an adequate ethical theory, providing ourselves with a "check list" for later use. These aspects of ethics involve (1) the *goals* for the good life that we should seek—the "material" of ethical interest—and (2) the *principles* by which we should rightly regulate our seeking—the "form" of ethical concern. Since we want theoretical coherence as well as adequacy (1.4), we shall have to pay attention also to the way in which these two aspects of ethics, the good and the right, go together.

a. What is Good?

Some things are sought by people for their own sakes, because they are intrinsically satisfying: e.g., being loved and loving, enjoying physical pleasures, exercising muscular or mental talents, having aesthetic or religious experiences, or the like. Other things, e.g., wealth, power, social status, etc., are sought because of what they can contribute to the attainment of the experiences that are intrinsically satisfying. The first kind can be called *intrinsic* goods and the second *extrinsic* goods. Some people have trouble telling these apart, especially since people are capable of becoming attached to extrinsic goods—making them, as it is said, "an end in themselves." But the distinction remains an important one, even though some intrinsic goods (like the enjoyment of health) may also be important conditions for other goods as well. Money, on the other hand, considered as a medium of exchange is entirely an extrinsic good. Those who allow themselves to be obsessed with such *mere* means may miss out on the ends for which they are meant to be used. As the old song says: "When you kiss a dollar bill, it doesn't kiss you back."

Both kinds, intrinsic and extrinsic, are "good" in a way that is neither "ethical" nor "unethical," "moral" nor "immoral."[2] Merely possessing them fails to carry any of that special type of approval that we reserve for moral praise and lacking them does not lead to moral blame. If, for example, some people are rich, powerful, loved, and pleasured, this fact does not by itself make them moral. And if others are wretchedly poor (like St. Francis) or in searing pain (like Jesus at his crucifixion), this does not make them wicked. Let us, with William Frankena, call them the *nonmoral* goods,[3] to distinguish them from anything either moral or immoral.

There must be some connection, however, between the non-moral goods and ethics. One indication of this connection is the widespread disapproval experienced when someone judged as wicked, or even not especially deserving, enjoys a dis-

---

[2]For the purposes of this discussion, "ethical" and "moral" will be taken as rough synonyms for one another.
[3]Frankena, *Ethics*, Chap. five.

proportionate share of these goods; another is the frequent and often intense concern felt when, as the Psalmist put it, "the righteous suffer."

If we ponder the basic nonmoral goods, we find that human beings may not, after all, be so radically different in what is satisfying.[4] At the basic level, prerequisite to all the other nonmoral goods, is healthy bodily survival (intrinsic good). This requires as means (extrinsic goods) food, water, and air as well as protection from predators—animals or other humans—and from the weather. Not only these bare minima, but also fuller material security for comfort (intrinsic good), and beyond this wealth (extrinsic goods), are sought. Along with these, people value society with others for friendship and love (intrinsic goods), and the various means thereto. Within the social framework, persons value recognition and status which give self-esteem (intrinsic good). Since persons are not all alike, individuality calls for freedom to exercise one's particular abilities and to participate actively in the choices that shape one's life—the freedom to exercise agency and, at best, enjoy personal creativity (intrinsic goods).

We could summarize the nonmoral goods as:

1. Survival
2. Health
3. Material security, comfort, wealth
4. Society
5. Individuality, agency, creativity

b. What is Right?

It seems plausible to argue now that ethical theory should recognize the above goods as generally and intensely valued by human beings (3.1.i). If so, it is not arbitrariness but simple realism to insist that such values should be fulfilled insofar as possible. Actions that lead to the attainment or enhancement of these goods, then, should be acknowledged as "right." At this point an *ethical* appraisal has entered, i.e., that it is morally approvable ("right") to act in ways that lead to intrinsic satisfactions. The converse is that it is morally condemnable ("wrong") to act in ways that diminish or frustrate such satisfactions, all other things being equal.

But such a principle runs immediately into problems. What, exactly, does it mean for "all other things" to be equal? Granting that it is morally right to enhance satisfactions in general, how should this be done if enhancing someone's satisfaction diminishes satisfaction for someone else? Likewise, *who shall count* as "someone"? Does my neighbor have as much status as "someone" as does a member of my own family? Does every person count as "someone" equally? If at this basic level of valuing there is no legitimate basis for discriminating among *human* valuers, why not push the argument further and argue that there is no basis, either, for dis-

---

[4] Abraham Maslow, *Motivation and Personality* (New York: Harper & Row, 1970).

criminating against other *animal* valuers for whom survival, health, security, their own society, and (perhaps) individuality are also basic goods?[5]

Such questions show the need for some principle of distribution of goods. If, as was said earlier, the concept of good is the "matter" for ethical theory, its "form" is fairness or justice. For ethical theory, like theory in general (1.4), only reasonable—consistent, coherent, and adequate—policies for seeking the good will be morally right. This means, at a minimum, that if we encounter conflicts among those who seek the basic nonmoral goods of life (and in real life we always encounter conflicts), the only times it would be morally reasonable to discriminate for or against any "someone" would be under circumstances when there are *relevant differences* to account for the difference in treatment. This follows from the basic rule of consistency. Under like circumstances it would be as inconsistent to draw different policies for action as under like premises it would be inconsistent to draw different conclusions for belief.

The formal Principle of Justice, then, requires that our actions, to be morally right, be non-discriminatory unless there is a justifying difference. That difference, in turn, has to be of a sort that would justify the proposed discrimination under any like circumstances. This does not mean that all actual treatment needs to be exactly alike. The world is full of relevant differences. If I insist that my three-year old child go to bed at 7:30 P.M., it does not follow that my thirteen-year old must do the same. The ten years' difference in age will involve relevant differences (in growth needs, in social capacities, in obligations such as homework) on which to make a distinction. But if I run a nursery, and I let some of my three-year olds play longer than the others merely because (say) they have blue eyes—or for no reason at all—I would be discriminating without a relevant difference and could be held to be unjust.

From this Principle of Justice as relevant equality, important consequences follow. It means that if I am dealing with the basic and universal goods of life, the fundamental "matter" of ethics, it is hard to see any relevant "formal" difference between the someone nearby who values them and the someone far away who similarly values them. My family and my neighbor are both to be considered equally, from an ethical point of view, unless there is some further relevant difference—besides merely being "family," an accident of biology, like being blue-eyed, that seems of little general relevance for justifying unequal treatment. There are normally, of course, genuinely relevant differences that can and do justify unequal treatment: I have obligations concerning my children, for their support and shelter, for example, that I do not have to neighbor's children, whom I did not bring into the world. But apart from such specifiable and justifying differences, the mere fact of family "propinquity" would not seem especially relevant.

"Who is my neighbor?" If this question is meant to ask, as it did in its New

---

[5]Peter Singer, *Animal Liberation* (New York: New York Review 1974). Tom Regan and Peter Singer, *Animal Rights and Human Obligations* (Englewood Cliffs, N.J.: Prentice-Hall, 1976). Richard Morris and Michael Fox, *On the Fifth Day: Animal Rights and Human Ethics* (Washington, D.C.: Acropolis, 1978).

Testament context,[6] who are the "someones" to whom the principles of ethical treatment apply, then the consequences of our line of reasoning can push us far beyond our physical and temporal neighborhoods, as in the famous biblical answer (the parable of the Good Samaritan) that this question provoked. Whoever is in need of the basic goods of life has a *prima facie* claim to them, all other things being equal. I have need for them; my immediate family has need for them; my geographically close neighbors have need for them; but also my geographically, culturally, and racially distant "neighbors" have need for them; my yet unborn, temporally distant "neighbors" have need for them; and perhaps the burden of proof is on those who would deny any consideration to members of other species, "neighbors" in fur and feathers who certainly differ from human beings, but perhaps do not differ wholly, or in wholly relevant respects, when it comes to at least some of the basic satisfactions of life.

c. What Ought to Be Done?

These considerations of the basic nonmoral goods and the basic obligations of justice need to be brought together in a theory of ethical obligation, since separately they pull in different directions. The quest merely to maximize the acknowledged goods of life is, by itself, heedless of the formal limitations of fairness. An ethic, in other words, based simply on the principle that one should always maximize the net sum of intrinsic satisfactions in one's world will, unless supplemented, be blind to the problem of how fairly those satisfactions are distributed. Piling up "the greatest (net) happiness for the greatest number" may be popular with the fortunate ones who reap the benefits; what shall we say about the forgotten ones who pay the costs?

On the other hand, an ethic that is only concerned with formal rightness and not at all with enlarging the basic goods of life may, unless it is brought back to earth, forget to acknowledge that it is precisely because the nonmoral goods are good that principles of equity in their distribution are needed at all. The formal aspect of ethics without material satisfactions is empty; the material aspect of ethics without formal guidance is blind.

How then can we summarize this theory of moral obligations? First, we are obliged to refrain from engaging in activities that will diminish the good in the world, or (all other things being equal) harm any "neighbor," in the large sense we have adopted. This, still close to Frankena,[7] we can call the *Principle of Nonmaleficence.* Second, we are obliged to contribute by our activities to the creation of good in the world and (all other things being equal) to helping any "neighbor" attain them, whomever or whenever that might be. Following Frankena, we can call this the *Principle of Beneficence.* The first corresponds to what is sometimes called the Silver Rule: "Do *not* do unto others what you would *not* have done to you."

[6]Luke: 10.
[7]Frankena, *Ethics*, pp. 45–48.

The second, more positive, corresponds to the more affirmative Golden Rule: "*Do unto others as you would that others should do unto you.*" Which rule is primary? Intuitions may differ on this, but it seems plausible that even more basic than creating new good is the obligation to be careful that one's actions do not leave the world or others worse off than before. The frequent complaint against "do-gooders" rests on this concern: that some, by insisting on acting to achieve what is believed would be new good, actually do harm.

Third, we are obliged to weigh our actions by the *Principle of Justice* that was implicit in the phrases, above, "all other things being equal." This means that to do right we need to do more than maximize basic nonmoral good, we need also to consider all who are interested in that good and to accept the burden of justification for any discriminations we make in their treatment.

"Do not destroy good." "Try to create good." "Be fair." These principles, together with the unpacking of what is implicit in the nature of some of the basic goods themselves, will be the standards by which we carry out our critique of the technologies of our modern world.

6.2. PROBLEMS AND POSSIBILITIES IN TECHNOLOGY ASSESSMENT

Ethical critique is not the only type of assessment, of course, that can be applied to contemporary technology. Other methods include *cost-benefit analysis* and *risk-benefit analysis.* In addition there is *alternatives assessment*, such as is sponsored in the United States by the Congressional Office of Technology Assessment. It is useful to contrast ethical critique with such other approaches to assessment. If we assess the assessors with the concerns of ethics in mind, what do we find?

a. Defining Goods and Harms

When a new technology, or a major application of an existing technology, is assessed, the first need is to determine what are the "goods" and "harms" involved and to define them in a way that allows responsible decision making. Since such nonmoral goods and harms are the "matter" of ethical theory, this is without doubt a vitally important question not only for policy makers but also for philosophers.

The most common approach today is to develop a cost-benefit analysis or, under special circumstances of potential harms to human life and health, a risk-benefit analysis. These analyses are aimed at clarifying the complex issues that always surround consideration of a significant technical impact on society or the environment. The "harms" or destructions of good are listed on one side as "costs" of the intervention; the "goods" are listed on the other side as "benefits"; each column is added, and the costs are subtracted from the benefits. If the costs are greater than the benefits, so that the "bottom line" is a negative number, the project is presumably not worth pursuing, but if the benefits outweigh the costs, so that the "bottom line" is positive, then presumably it is reasonable to go ahead.

Using cost-benefit and risk-benefit analysis is an alternative to blind decision making. It gives those who think about technologies and their impact a regular method for settling questions based on calculation instead of hunch. It is a way for those who are accountable to give an account. It is a method that is well entrenched. Business and political leaders have come to depend heavily on it.[8]

These analyses, however, seen from a comprehensive philosophical point of view, have significant problems in defining the nonmoral good. One difficulty is forced by the need to define the goods and harms involved in terms that can be quantified, added and subtracted, usually in terms of money. This quantification and monetization of the valued and disvalued aspects of an issue are essential to the analysis. Without it one could never reach a "bottom line" at all. But clearly some valued goods are extremely difficult to translate into any quantitative terms at all, least of all money. Exactly "how many times more beautiful" is this landscape than that one—or, harder yet, than a certain seascape or sunset? The question, even when comparing similar aesthetic objects, is baffling at best and absurd at worst. This makes it even stranger to ask how much more or less such a landscape is worth in dollars than the tall smokestack being planned for its center. Hence it is tempting to ignore the value of intangibles—beauty or ugliness, loyalties to neighborhoods, historical significance, etc.—when drawing up cost-benefit analyses, thereby giving a heavily disproportionate weight to those features of the situation that lend themselves more easily to the method. *Bias in favor of the measurable and marketable* is built into the approach.

In an attempt to give intangible values some recognition, cost-benefit analyses sometimes attempt to create a shadow "market" by asking people in surveys what they would be willing to pay for some good or to avoid some harm. How much would you be willing to pay (per year?) to avoid having an oil refinery built across the road? How much to keep the landscape clear of the smokestack? How much to continue being able to enjoy such sunsets? But this method suffers from epistemological difficulties, even if people are willing to put a price on such preferences. One difficulty is the problem of knowing how accurate such shadow pricings might be without any real need or opportunity to pay. I might say that I would "sell my shirt" for some value in the abstract, but it might turn out that I would actually have more pressing demands on my budget; or I might tell the survey taker that I would not "give a nickel" for the view from my window, but when the time came I might devote heavy resources to fighting the smokestack. A cost-benefit analysis based on the inclusion of such soft data is hardly likely to be as hard-headed—or ethically responsible—as it seems.

Another epistemological difficulty is in knowing whether future preferences will be the same as present ones. As Robert Coburn writes, "Policy decisions as regards the development and diffusion of technologies, like personal decisions about important matters such as whom to marry and what career to follow, can have long-range and profound effects on the values and hence preference-structures

[8]Richard J. Mishan, *Cost-Benefit Analysis: An Introduction* (New York: Praeger, 1976).

of those affected by these decisions."[9] But if this is so, any major decision will influence future values and change them in unpredictable ways.

Still more epistemological puzzles are caused by attempting to know even one's own preferences when dealing with imponderables such as those often appearing in related risk-benefit analyses. How much would it be worth to me today to reduce by .005 percent the chance of cancer in my thyroid gland twenty years hence? This need not be put into monetary terms. Would it be worth giving up drinking milk for a week after a nuclear "incident" in a neighboring state? Would it be worth going on my next business trip by train to avoid high altitude radiation from cosmic rays? Is it worth a great deal more than this to me? Is it not worth anything at all? How can ethically sound decisions be made concerning the good for others if I am unable to define even my own good more clearly than this?

Dealing with future values as contrasted to present ones introduces another element of bias into cost- and risk-benefit analyses. Future value is always less secure than present value, and is therefore subject to systematic underestimation. The clearest example of this is the way in which the discount rate tends to shrink the value of future dollars in proportion to the interest rate at any given time. The higher the compound interest rate, the earlier the date by which tomorrow's dollars will shrink exponentially toward negligible worth. Quantification and monetization of goods and harms in cost- and risk-benefit analysis tend, accordingly, to *bias decisions toward present exploitation of opportunities and against the values of conservation for the future.*

A final ethical difficulty that combines several of the above is the problem, in cost- and risk-benefit analyses, of defining the value of human lives. This issue is most often handled by monetizing the lives concerned in terms, say, of expected lifetime earnings. There are profound conceptual inadequacies in this approach, however. Ian G. Barbour summarizes these well:

If applied consistently, the method would require that the lives of the elderly would be considered valueless. If future earnings are discounted, a child's life would be worth much less than an adult's. . . . I would maintain that there are distinctive characteristics of human life that should make us hesitant to treat it as if it were a commodity on the market. Life cannot be transferred and its loss to a person is irreversible and irreplaceable.[10]

This does not mean, on the other hand, that human life must be "priced" at infinite value, since risks to human life cannot be avoided and such an extreme "pricing" policy would rule out all realistic thought about actions or inactions, technological or not, that might hazard a human being in the slightest degree. The problem is with the method of "pricing" itself. As Barbour concludes, "The cost in

[9] Robert Coburn, "Technology Assessment, Human Good, and Freedom," in *Ethics and the Problems of the 21st Century*, eds. Kenneth E. Goodpaster and Kenneth M. Sayre (South Bend, Ind.: University of Notre Dame Press, 1979), p. 113. By permission of Notre Dame Press.

[10] Ian G. Barbour, *Technology, Environment, and Human Values* (New York: Praeger, 1980), p. 173.

human lives should not be aggregated with economic costs, but kept as a separate kind of cost concerning which accountable decision makers, rather than technical analysts, should make the inescapable value judgments."[11]

In sum, technology assessments that rest mainly on traditional cost- and risk-benefit calculations may be ethically suspect if they give systematic preference, in the basic process of defining relevant nonmoral goods, to *quantitative, marketable, immediate* aspects of the situation to be assessed. An adequate ethical theory will recognize these aspects as genuine and important, of course, but it will recognize them as aspects rather than the whole. The "fallacy of misplaced concreteness,"[12] the temptation to treat a significant but partial abstraction as though it were the real thing, combines here with the power of successful methods to perpetuate themselves (3.3 and 4.3) and to oppose speculations about their limitations. But ethical theory needs nevertheless to exercise its gadfly role, insisting on the importance of the qualitative, intangible, and imponderable aspects of the good as well.

b. Applying the Principle of Justice

If the "matter" of ethical concern is the nonmoral good, its "form" is fairness, equity, or justice. How does traditional cost- and risk-benefit analysis fare when assessed by this ethical standard?

Cost-benefit analysis is not designed to pay attention to the ethically crucial question: "Who pays the costs, who gets the benefit?" Typically, such analysis reaches its "bottom line" by aggregating the costs and the benefits, then subtracting the one from the other. Its goal is to determine, within its limited definition of the goods and harms involved, the *net* good or harm that a technological intervention will produce. If the smokestack in my neighborhood can be calculated to allow millions of distant persons to benefit from slightly cheaper electricity bills, this can be multiplied (thanks to the large numbers involved) into a major benefit; if the smokestack results in higher cleaning bills, greater medical costs, and residential property devaluation for my immediate neighborhood, this can be calculated as well and subtracted from the "greater good." The net benefit, let us assume, is positive, and the decision is clear. Or is it? The Principle of Beneficence, to create greater good, is satisfied, but the Principle of Justice has been overlooked. Equal treatment is not being provided to equally "deserving" parties. Why should the accidents of geographical location warrant my bearing the bulk of the costs of this technological intervention while others, not relevantly different from me, reap the benefits? It is unfair. At the minimum, in a just world, I would deserve compensation from the benefiting majority for my unwarranted costs. They should pay the extra part of my cleaning bills, subsidize the costs of replacing my corroding car more often, help me with my doctor's bills. . . . But the world approved by cost-

---

[11] *Ibid.*, p. 173.

[12] Alfred North Whitehead, *Science and the Modern World* (New York: The Free Press, 1929, paperback edition, 1967), p. 51. See also 5.4 above.

benefit analysis is not a just one. How could it be, since the "bottom line" for me might be the enjoyment of health and life itself (6.1.a). The mortality rates for my children and me cannot be "subsidized," even if the diffuse and normally heedless majority of my fellow-citizens or the power company should decide to try.

Distributive justice is not even a goal of cost-benefit decision making. It must be tacked on as an afterthought if it is considered at all. Often it has not been considered in making significant technological decisions, though in some famous cases in recent experience those considerations have been voiced. The decision of the United States Congress not to proceed with the civilian Supersonic Transport project, for example, was at least in part related to the argument that a wealthy few travelers would gain all the benefits while large numbers on the ground would pay the costs in noise pollution, cracked window panes, and (possibly) cancers caused by damage to the ionosphere. More often, however, the cost-payers are the poor and relatively powerless members of society, whose burdens in polluted and little-visited neighborhoods can be easily buried in the "reasonable and objective" "bottom lines" of cost-benefit calculations. Risk-benefit analysis, since it deals with issues of human life and health, often is somewhat more attentive to the question of who bears the risks for the benefits of technologies; but even so it may fairly be concluded that the aggregating methods of cost- and risk-benefit analysis *tend to be defective in concern for distributive justice.*

Justice across geographical and economic lines is one important matter for ethical consideration; *justice across time* is another. Why, just because I happen to be born into a certain generation, should I be discriminated against compared to those born into different generations? We have already seen that the bias of cost- and risk-benefit analysis is toward favoring short-range exploitation of opportunities and resources. In this bias lies a temptation to neglect what Barbour calls "intergenerational justice."[13]

Technological decisions often have extremely long-term consequences, especially regarding the extraction of non-renewable resources and the depositing of toxic substances (6.3.c). Ethical adequacy would require that in fairness the risks and the costs of all such decisions be considered over the full-time frame of the impact. Future persons cannot speak for themselves. They do not vote. There are even conceptual problems about whether "they" can have rights at all, since "they" are only probable and not actual people, abstract and not specifically identifiable. But whether our ethical obligations are described in the language of respect for rights, or in some other way, the ethical assessment of technology will require a long time frame and a respect for the principle that future persons, once they become actual persons, will have no less claim on equitable treatment by us than persons living today. If cost- and risk-benefit analysis minimizes this significant ethical concern, we may conclude that it *tends to be inadequately concerned for intergenerational justice.*

Are human persons, present or future, the only objects of ethical concern? The

---

[13]Barbour, *Technology, Environment, and Human Values*, pp. 84–88.

bias we have noted toward the monetizable and the marketable in cost- and risk-benefit analysis tends to reinforce its *bias toward the anthropocentric* definition of values. Animals do not bid in the marketplace, they are bid upon. Therefore, for cost-benefit calculations, their entire worth is defined as commodities. But even if this is so for such methods, it need not be so for ethical theory. Animals have a life apart from the aspect of their existence as marketable commodities. It would be a clear instance of the fallacy of misplaced concreteness to identify the whole value of an animal's existence with its price tag. The "more" of animal existence plausibly involves some capacity for experiencing the environment, for pleasures and pains, for aims and satisfactions and frustrations. These capacities may be different from their human homologues and may differ greatly from species to species. A horse and an oyster should not be equated any more than a horse and a human should be absolutely set apart in capacities.

If this is so, and if the Principle of Justice requires that discriminations of treatment be proportioned to relevant and justifying differences, then some animals may, depending on their capacities, be included among the "someones" whose good or harm is of proper ethical concern. At least the inflicting of random or unwarranted pain on creatures capable of feeling pain would be a violation of the Principle of Non-maleficence. Given that much, we may conclude that non-human interests need to be taken, *to some appropriate degree* for their own sakes, into ethical consideration—which marks the end, in principle, of anthropocentrism. If we venture into this new ethical territory, we have left the thought-world of the marketplace far behind. Our ethical assessments of technologies will now need to include the question whether aspects of nature as well as human society will be harmed or helped by some technological application. Our views on how much within nature deserves to be given ethical respect will depend on our metaphysics. One more conclusion may be drawn about cost- and risk-benefit analyses: by their anthropocentrism they *tend to ignore considerations of ecological justice.*

It seems clear that cost- and risk-benefit analyses by themselves are no substitute for careful ethical thinking about technology. Overreliance on their use might indeed be ethically inferior to decision by "mere human hunch" since, left to themselves and followed too literally, they can distort values and lead to injustices. But if used with caution and with constant acknowledgment of larger governing standards, their limitations may be kept under control. As Barbour notes, "These limitations are not serious when a project has narrow and clearly defined objectives, when the main impacts are physical and readily quantifiable, and when there is a small number of options for achieving the objective."[14]

c. Seeking Responsible Agency

For larger questions, the approach taken by the Office of Technology Assessment (OTA) of the United States Congress is, from an ethical standpoint, more promising.

[14]*Ibid.*, p. 171.

This does not mean that the record is perfect. The OTA is, after all, a human political entity with the weaknesses and pressures that all flesh and politics are heir to. But at least an earnest attempt has often been made to keep cost- and risk-benefit calculations in the wider context of significant but intangible social goods and standards, including beauty and distributive justice. Its multi-dimensional examination of Coastal Effects of Offshore Energy Systems,[15] for example, not only showed sensitivity to a broad range of possible goods and harms, and to the social distribution of the corresponding risks and costs, but also ventured to speculate with thoughtful imagination on alternative technological responses to the given problem.

Better yet, the process is open to the public, both in its reports and in its methods of gathering material to go into those reports. One fundamental "good" to be sought by the Principle of Beneficence, we remember (6.1.a), is the positive freedom to exercise control over our lives, the freedom of responsible agency. To the extent that technology policy in a society is set without the participation of the ordinary members of that society, the good of responsible agency is diminished for them. The Principle of Non-maleficence is violated. To the extent, conversely, that citizen voices can have effective say in the assessment and determination of which technologies they live with, the ethical situation is improved, even apart from the substance of the decision itself.

Among the relevant and justifying considerations in judging the justice of distributing harms along with benefits is the degree to which the risk of harm is or is not voluntary. If some people choose to smoke cigarettes, despite the clear evidence that it is harmful to their health, and if they later contract cancers as a result, that gives rise to one ethical situation. It is deeply unfortunate, we may conclude, but not wholly unjust. "They asked for it." The ethical situation, however, is different when their non-smoking spouses (for example) contract smoke-related cancers. That is not only unfortunate, it is unfair. But since marriage is voluntary, the ethical situation is even sharper when strangers who happen merely to share the same workplace come down with diseases caused by someone else's habit. That is outrageous.

This sensed moral relevance of voluntary risk is reflected in the public willingness to acknowledge and accept the hazards of travel by automobile—we have exercised our responsible agency when we choose to take the highway—while often showing much less willingness to accept the statistically smaller hazards of nuclear energy production. Being poisoned by someone else's faulty design or lapse of attention half-way around the world is morally less acceptable, more unfair, than being smashed by my own, or even someone else's, carelessness at the wheel. At least I *chose* to drive, knowing the risks. Ethically it makes a relevant difference if consent was given and personal autonomy respected.

For this reason, and for the others mentioned, OTA-type assessments, including

[15]Office of Technology Assessment, *Coastal Effects of Offshore Energy Systems* (Washington, D.C.: OTA, 1976).

participation of citizens in the process of making vital public decisions affecting survival, health, physical security, and the character of one's society, are ethically praiseworthy. Such participation enhances the nonmoral goods of personal individuality and exercise of agency while it alters the circumstances and allows for increasing justice. For similar reasons even "private" decisions on technology, when they have profound effects on persons now living or in future generations and when they have significant impact on the environment, are rightly subject to citizen participation. Just as your freedom to smoke (properly) ends where my lungs begin, so an electric utility's plan to erect a smokestack ceases to be merely private where my view, property values, and health are affected.

We have been considering ideals, of course. The realities are always more disappointing. An adequate ethical critique would need to examine the extent to which genuine citizen participation is facilitated in the actual political process. How representative are our representatives? Is the Advisory Council of the OTA, for example, in need of more minority, poor, labor, and consumer voices? White, middle class, mainly technically oriented males may attempt in good faith to speak for all the rest, but it is difficult at best and ideally should not be necessary at all. Does government in any parliamentary democracy properly listen to the concerns that finally filter to it through whatever processes have been devised, or do most governments respond to other short-range pressures, electoral and economic? Is the "revolving door" between economic interest groups and their public regulators too strong for responsible citizen agency to be real? These and other questions need to be asked and answered in any full ethical critique of the methods by which modern societies deal with their technologies. Here technology blends into the fabric of society itself, just as ethics blends into political action.

6.3. ETHICAL QUESTIONS FOR CURRENT TECHNOLOGIES

Controversy over specific major forms of contemporary technologies may be expected to continue and intensify as the twentieth century draws to its turbulent end. In this chapter we cannot hope to anticipate the direction of future debates, much less offer neat solutions to the problems of our era. But a wise use of ethical theory may be able to cut through rhetoric and confusion by insisting on attention to a few ethical fundamentals. Let the details be supplied as appropriate to the issues. It will always be appropriate—and important—to ask: what are the nonmoral goods and harms involved, and what are the considerations of justice?

a. Automation: Technology in the Workplace

In a broad sense, industrial technologies since the introduction of mass production and the conveyor belt have been steadily trending toward the automatic. The major nonmoral good promised by these industrial technologies is material security, in the sense of vastly increased productivity per unit of human labor. It is the genuine good of Fuller's "doing more with less" (5.3). The necessities and comforts of life have never in history been provided so bountifully to so many.

What are the harms that should concern us? One frequent complaint voiced by those who work with industrial technology is that they lose thereby the good of their own (intrinsically valued) personal autonomy. Ruled by the inexorable demands of the machinery or the production line, their sense of responsible agency is eroded so much that basic human dignities are lost. The image of Charlie Chaplin, in *Modern Times*, struggling to keep up with the ever-accelerated conveyor belt and finally, wrench in hand, going berserk from stress, vividly captures this sense of harm. The feeling of losing control, of being captured or even "swallowed" by the machine (these images are featured, too, in such films as *Modern Times* or Fritz Lang's *Metropolis*), are intrinsically painful ones.

In craft technologies, the skilled worker's sense of creativity, pride in agency, autonomy, is able to be expressed. In automated technologies, these valuable satisfactions tend to be replaced by a sense of dependency, partial exercise of abilities, and personal replaceability. Thus a second basic harm is the danger of *loss in experienced individuality and self-esteem.*

Can automation in the workplace avoid or minimize these harms while retaining the goods it provides? This is a key issue for an ethical assessment of industrial technology, but the answer is not clear. Attempts have been made, in Sweden and elsewhere, to bring a sense of autonomy, craft, and wholeness into the production process, but the effectiveness of these arrangements remains uncertain. As the trends toward more fully automatic systems, including robotics, continue, however, it is at least possible that Charlie Chaplin's "mindless" tasks will be increasingly lifted from human workers and that greater responsibility and dignity will be required as the distinction between worker and manager becomes more and more blurred. It is, on the other hand, also possible that Marcuse's warnings (5.5) of the flattening of spirit that comes from homogeneity will be confirmed. At any rate, the Principle of Beneficence would place a high priority on eliminating or minimizing the nonmoral harms of automation and working toward adding further varieties of vital but intangible human goods to the material goods we have come to expect.

To discuss workers and managers raises, in one of its forms, the problem of *justice.* Are there relevant and justifying reasons for the distinctions of fulfillment, dignity, physical and psychological well-being, and material reward between the different persons involved in the production process? If education and training are given as justifying reasons, are there also justifying reasons for the differences in the ready availability of such education and training to all the persons concerned?

Again, to reopen Marx's question (5.2), considering the great cost and potency of contemporary productive technologies, is it equitable that some but not all persons can claim ownership of, and limit access to, these vital means of production? Aside from questions of ownership, is it fair to replace workers with machines if the result of greater ease and dignity for some means unemployment for others? It may, of course, turn out that automation does not undermine employment. It may only change the kinds of jobs that must be done.[16] But in any event, the ethical assessment of technology in the workplace will require that dislocations be compensated and that retraining be provided for those who lose their accustomed jobs. And the ethics of the situation would require an answer in case employment itself is reduced. What is the fair way of responding? Shall the remaining employment be spread out by everyone working less, or shall the unemployed be subsidized out of the surplus product gained for society by new technologies? The one answer ruled out by an ethical assessment would be that society do nothing for those who, through no relevant fault, suffer from technological displacements.

b. Computers: Technology and Electronic Intelligence

The possibilities of robotics and increased automation in the workplace are dependent to a large degree on the technologies of computerization. Sharp debates may be expected to continue over the computer and its place in society.

What are the nonmoral goods brought by the computer? The answer is many-faceted, but if the bounty of industrial technology is increasingly dependent on computers, an appropriate share of the credit for this bounty in material security should be acknowledged for computers from the start.

Beyond this, the computer makes possible the organization of complex systems— the inventories, mailing lists, transfers, billings, orders, police records, etc.—that make modern large-scale society possible. There is no need to exaggerate the dependence of complex enterprises and social organizations on computers (e.g., the atomic bomb was invented and World War II was fought without them); but entirely without exaggeration it is obvious to all who live in the contemporary world how pervasive and essential the computer has now become to our way of maintaining current civilization. The clear trend is to enmesh our lives ever more closely with electronic intelligence.

Still further, computers serve the goods of health in diagnosis and hospital care. They also serve the creative values, like the quest for understanding in science and the pursuit of beauty in the fine arts. In countless ways they are being used to enhance individuality, thanks to the spread of personal computers. Some nonmoral good is found in sheer play with computers, at various levels of sophistication.

What are the harms that worry critics? Expressed ethically, a serious charge is that computers deprive persons of individuality by invading privacy and by facilitating social centralization. Privacy is an extrinsic nonmoral good that may have many important consequences for intrinsic goods. It may allow love to be expressed and friendship to blossom without the inhibitions of curious or disapproving eyes. It gives individuality a chance to develop and mature apart from the surveillance and control of dominant orthodoxies. It provides scope for political activity—the exercise of responsible autonomy—even when the aim of that activity is to "throw

[16]Emmanuel G. Mesthene, "Technology as Evil: Fear or Lamentation?" in *Research in Philosophy And Technology*, ed. Paul T. Durbin (Greenwich, Conn.; JAI press, Vol. 7, 1984), pp. 60–61.

the rascals out." If the speed and retentiveness of modern computer technology—given electronic record-keeping of bank accounts, telephone records, employment history, library book borrowings, police records, census information, etc.—are capable of tearing away these protections by allowing the would-be controllers, or the merely curious, to gain access to information (and patterns in the information) that would otherwise be unavailable, then profound harms may be done to all these precious goods.

Another harm that most have experienced directly is the sense of helplessness and indignity that comes from attempting to "argue" with a computer. Such *lost sense of responsible agency* is even more seriously a problem when computers are required to make decisions of great consequence, as in diagnosing a poison and prescribing an antidote or in determining whether to launch missiles, before human decision makers, in either case, would have time to work out conclusions from complex data. The chilling sense of having one's fate taken from one's hands, to have the machines take over the helm of human destiny, is a profoundly felt disvalue, whether the machines will do better than the humans or not.

Issues of justice raised by computers range from tort law over who is rightly to be held liable for errors made in medical diagnosis due to a software defect, on the one hand,[17] to major social questions of the distribution of expertise and power, on the other. To look briefly at the latter, it may be that ethical distinctions will need to be made between the main frame and the personal computer. The former tends (it could be argued) toward centralization and the increasing monopolization of technical expertise and thus poses problems of distributive justice as well as the nonmoral harms of felt powerlessness, lost privacy, and standardization. The latter tends toward decentralization and the democratization of computer know-how, thus to distributive justice, and to new nonmoral goods in individuality, creativity, and self-esteem. These issues are far more complex, of course, especially since large computers can be used in "networking" among many users and tasks, and since personal computers depend on large manufacturing and service facilities, but it is helpful to see that significant ethical distinctions can be made even within a technology.

c. Nuclear Energy: Technology in Peace and War

No technologies in our time are more controversial and ethically arresting than the nuclear ones, whether they are designed for civilian power generation or for military purposes. The absoluteness of the forces concerned and the consequences of their use contribute to the utmost seriousness with which they must be treated and discussed.

The goods foreseen by these technologies can be strongly stated. The development and use of the first atomic bomb was perceived in its time as a matter of

---

[17]Vincent M. Brannigan, "Liability for Personal Injury Caused by Defective Medical Computer Programs," in *Ethical Issues in the Use of Computers*, eds. Deborah G. Johnson and John W. Snapper (Belmont, Calif.: Wadsworth, 1986).

survival, against the dangerous hostile forces who were also engaged in its pursuit.[18] The subsequent transformation of nuclear fission into "the peaceful atom," as President Eisenhower called it, was perceived as a means to unlimited and cheap energy for the sake of material security and comfort. These two fundamental values remain the primary ones. Now "survival" is sought through mutual deterrence in which, it is argued, unilateral abandonment of nuclear weapons would open the way to self-destruction; and "material security" is defined with increasing awareness of the historically short time, measured realistically in decades, before the earth's remaining oil reserves will be effectively exhausted. Few love the atom for its own sake, but many see in it the only alternative to the enormous harms of extinction or economic collapse.

On the other hand, the possible harms posed by nuclear technologies are no less apocalyptic. The nightmare possibility at any time of a full-scale nuclear exchange between hair-trigger superpowers raises again the threat to survival in its most ghastly form. There is at this writing good evidence that this threat, due to "nuclear winter," could extend to both hemispheres and to all civilized human life, as well as to the extinction of other species due to radiation sickness and temperature and climate modifications, following blast effects. The chance that an atomic exchange, once begun, could be "limited" to less than this is highly problematic. Even if it could, the destruction and sheer suffering from a "limited" use of hydrogen bombs would likely be greater than any previous horror in human history, including the horrors of Hiroshima and Nagasaki, whose explosions would now be dwarfed in comparison.

Peaceful nuclear power plants, despite official assurances of safety that the public has come (with reason) to distrust, are subject to accidents, small and large. The harms here are to health and material security of the most basic kind. Even at large distances, even across sovereign national frontiers, crops are ruined, citizens are terrified, and prospects for long and healthy life are threatened when catastrophes that "can't happen here" *do* happen. Some of the toxic materials are extremely poisonous and some are extremely long lasting. Even if not accidentally released, many of the radioactive substances created by nuclear reactions will threaten life forms for thousands of years. The problems of waste disposal with assured safety for survival and health, despite promises from experts and politicians, have not been solved; but new wastes pile up at an accelerating rate at temporary sites. These wastes already exist, whether we are ready or not, and seem, ethically considered, to demand not a mere emotional anti-technological response but a creative blend of theoretical and practical reason that will develop new "counter-technologies" for waste management, for extremely long-term social and environmental protection, and for permanent disposal. Some of the by-products of certain nuclear processes, particularly plutonium, are valuable for bombs and must be protected against terrorists who may feel that they have "nothing to lose," or against theft by other nations seeking military advantage.

[18] Arthur Holly Compton, *Atomic Quest* (N.Y.: Oxford University Press, 1956).

The issues of justice are no less pressing. The mind-numbing implications of atomic war for ecological and intergenerational justice are horrendous. Decisions made by members of our generation could irreversibly damage the biota on this precious planet for all time to come. Short of nuclear war, the catastrophes of nuclear accidents and the silent radiations from nuclear wastes may greatly harm future generations as a direct result of our hunger for plentiful electrical power and material comfort. The unfairness between species and generations is plain.

Back on the contemporary social level, what is sometimes overlooked in the technical arguments over the risks and benefits of nuclear energy is *justice in the distribution of the political authority to determine what our choices as a society shall be*. Could—and should—the citizens of contemporary parliamentary democracies, if a significant majority should prefer, effectively choose the risks of stopping the escalation of nuclear weapons instead of the risks of the balance of terror? Could—and should—the public, if sufficiently convinced of the dangers in prior "Faustian bargains" over nuclear energy plants, put a stop to the economic and political momentum that these technologies have accumulated? Ought ordinary people, inexpert in technical details, be allowed to choose the option of publicly supported research into alternative energy sources and, if necessary, to adopt a lower standard of energy consumption, even to accept severe economic hardship, for the sake of such values as health and ecological justice? Should they, to the contrary, be able to insist on building up nuclear deterrence and nuclear plants for electricity, if they prefer these despite the potential hazards, just as smokers are allowed to choose what they know is likely to damage them and those around them? How much social paternalism by the "experts" is ethically warranted, and how much individual self-determination? These are ethical questions not only about the fair distribution of political power but also about basic intrinsic goods of self-determination and responsible agency. They deserve pondering. They will be with us for a long time—if we are lucky enough to survive our answers.

d. Development: Technology and the Third Word

Transferring technology—or not—to developing nations, and if so what kinds of technologies, is another debate with deep human import. The good in view is clear and often urgent. Traditional societies around the world, typically found in the Southern hemisphere, are often grindingly poor. Modern technologies promise survival and health for starving, disease-ridden multitudes and material security beyond anything indigenous traditional technologies can provide. Besides the humanitarian goods, the economic development of a traditional culture creates a new market for the industries and enterprises of the developed world, thereby encouraging the creation of additional wealth for the providers as well as well-being for the providees. The Principle of Beneficence would seem to support, in general, the sharing of modern technologies beyond the cultures where they were developed.

On the other hand, the Principle of Non-maleficence would caution against doing good that might cause even more harm. One harm regularly overlooked in the economic calculations is the harm to the intangible aspects of the cultures to which modern technology is transferred. We have seen that modern technology is the embodiment or implementation of a specific way of thinking (4.3 and 4.4) with its characteristic set of explanatory categories and implicit attitudes toward nature and society. When modern technology is transferred, therefore, more is provided than hardware alone. The underlying fabric of the receiving culture—its ethical views, its patterns of approach to the world, its religious convictions—are implicitly jeopardized to the extent that the artifacts and methods of modern civilization entwine themselves within customs and institutions sprung from different mentalities. An ethical assessment will ask whether this is a justifiable intrusion. Are the implications of what comes subtly along with orders of spare parts and maintenance manuals fully understood and accepted by the receiving peoples? If not, is technology transfer in fact a form of cultural imperialism, a disguised secular missionary movement on behalf of the gospel of modern science and Western civilization?

Other causes for concern are more physical and obvious. Building factories near the cities in third world countries, together with short-sighted agricultural policies, have drawn or driven poor people away from the land, often in futile search for too-few available jobs, and thus has caused massive dislocation problems and huge, pestilent encampments of people even more desperately poor than they were before. Some have benefited, the rich most of all. But the Principle of Justice has often been ignored in the distribution of the additional wealth that technology transfer has created.[19]

The most visible harm from transfers of sanitation, public health, and energy-intensive agricultural technologies is the burden of overpopulation in the developing nations, resulting from the successful limitation of death rates without corresponding limitations in birth rates. From too many people come the further agonizing harms of pollution, natural resource depletion, pathologies of crowding, sickness, starvation, civil disorders, and threats of war. Should the Principle of Non-maleficence here override the Principle of Benevolence? Should transfers of modern technology and other forms of assistance to the poverty areas of the world be discontinued or significantly changed?

Garrett Hardin has argued bluntly that aid is futile. Poor populations tend to outstrip resources. The more poor people, the more suffering in the world. The developed nations do not have infinite resources, and, like survivors in a lifeboat with limited rations, should look to their own needs. If the desperate and drowning were all to be taken aboard, the lifeboat would capsize and all would be lost: "perfect justice, perfect disaster."[20] If modern technology transfers save lives,

---

[19]Frederick Ferré and Rita Mataragnon, eds., *God and Global Justice* (New York: Paragon House, 1985).

[20]Garrett Hardin, "Lifeboat Ethics: The Case Against Helping the Poor," *Psychology Today*, September 1974, pp. 38–43, 123–24.

according to Hardin, that may not be a good thing, after all, since "the life we save today breeds more lives in need of saving tomorrow."[21]

The "lifeboat" approach is rejected by many others, who look instead to finer discriminations and reforms in technological transfer policies. We in the developed world are not down to our last rations. We can afford, and by the Principle of Justice we ought to afford, to cut deeply into our luxuries for the sake of those who, from no morally justifying reason, or for reasons that have brought us unfair benefits in the past, are in desperate need.[22] Nor are we in a boat that can simply row away from the less fortunate. We are all "in the same boat," in an interconnected world where maldistribution can bring retribution as well as moral blame. Thus Lester Brown urges concentration on long-term methods of assistance, including stress on the education of women, whose expanded awareness (an intrinsic good of its own) could help to curb the birthrate both through delayed marriage and greater acceptance of birth control technologies.[23] E.F. Schumacher[24] and Ivan Illich[25] both argue for changes in the scale and quality of technologies, especially those designed to fit the human needs of the developing nations, while Erich Fromm[26] offers a psychological basis for redesigning the technologies of present "overdeveloped" civilization to enhance the human values, especially those of responsible agency, in our society as well.

Of course if one simply cannot do something, it makes no sense to say that one ought to do it. For such ethical prescriptions as these latter to be possible, we would need to turn our civilization toward significantly different thought patterns, since artifacts and techniques are embodiments of mentality. Therefore, if we are to take seriously such suggestions of "alternative technologies," for others or for ourselves, we need eventually to examine the prospect of "alternative techno-scientific thinking" as well. This is something to which we shall return (8.5).

e. Genetic Engineering: Technology and Artificial Life

The phrase "artificial life" may strike one at first as odd; but recalling our early reflections on the "natural" and the "artificial" (2.8), that is exactly what we should call life forms that exist only through the deliberate intervention of intelligence ("art") in their originating causes. Biological technologies, especially those of

[21]Garrett Hardin, *Promethean Ethics: Living with Death, Competition, and Triage* (Seattle, Wash.: University of Washington Press, 1980), p. 65.

[22]Peter Singer, "Famine, Affluence, and Morality," in *Moral Problems*, ed. James Rachels (New York: Harper & Row, 3rd edition, 1979), pp. 263-78.

[23]Lester R. Brown, with Erik P. Eckholm, *By Bread Alone* (New York: Praeger, 1974).

[24]E.F. Schumacher, *Small is Beautiful: Economics as if People Mattered* (New York: Harper & Row, 1973).

[25]Ivan Illich, *Tools for Conviviality* (New York: Harper & Row, 1973).

[26]Erich Fromm, *The Revolution of Hope: Toward a Humanized Technology* (New York: Harper & Row, 1968).

genetic engineering, are of great and increasing importance. What are some of the key questions for ethical assessment?

The goods sought by genetic manipulation of the codes by which living things transfer their traits include, among others, the basic value of health. First, the capacity to engineer the production of substances such as insulin, otherwise available only from slaughtered animals in restricted quantities, earns for genetic engineers the gratitude of diabetics. Children deficient in human growth hormone may now have access to engineered supplies, formerly restricted by the need to extract it from the pituitaries of cadavers. For some of these survival, too, depends on the availability of these organic substances. Second, birth defects may now be discovered before birth; and beyond this there are hopes for their prevention by means other than abortion, e.g., by gene therapy, replacing defective genes and thus preventing or ameliorating genetically caused defects. Third, applied to agriculture, the realm of possibilities for genetic engineering is practically unlimited. Perhaps a variety of corn can be developed which can generate its own insecticides, for example, thus eliminating the need for externally applied poisons. The development of vaccines, both for human health and agricultural applications, is also an anticipated benefit of genetic engineering. Thus the goods of survival, health, and material security are much in prospect from artificial life forms in a needy world.

The ethical assessment of harms, on the other hand, is not so obvious, aside from the much-discussed risk of catastrophic plague due to the release of artificial disease germs for which there are no natural defenses. Those obvious dangers have concerned genetic scientists themselves; and after an unprecedented self-imposed moratorium on research in this area,[27] the profession has satisfied itself that its precautions are adequate against these risks. Whether this judgment is correct, or whether it was unduly influenced by many pressures—career ambition, economic opportunity, international competition, and the like—only time can tell, though the matter is of great ethical as well as practical interest.

Other ethical concerns may range from questioning whether life itself is something that simply ought not to be manipulated, something that should stand outside the all-pervasive technological "Enframing" that Heidegger warns of (5.4), to warning that the creation and even patenting of artificial life forms will have the long-term effect of reducing respect for natural life forms, including human life. Beyond all this hangs the question: Should the human species itself be engineered to someone's specifications? Some may ask how our present generation has the right to determine what future generations shall be like. Others may ask whether we have the right to refrain. Whose preferences should rule: the reformer's or the conservative's? Should all the genes in the present human gene pool, including all the defects, be carried forward indefinitely merely because they are "there"? Is there something normative about the "natural"? Do we have the moral right, if we can prevent

[27]See the account, with accompanying citations, in Frederick Ferré, *Shaping the Future: Resources for the Post-Modern World* (New York: Harper & Row, 1976), pp. 54–59.

disease and deformity, increase health, and prolong survival, to withhold such goods?

These questions and many others must be asked as the infant biological technologies mature in the years to come. As always, the power to do something is double-edged. Where there was no possibility of designing the human race, there was no moral problem. The embodiment of techno-scientific thought in new theory-based biological technologies, however, has created unprecedented possibilities for action. Thereby it has posed new inescapable ethical dilemmas, since *to do* or to *refrain from doing* are both now becoming morally significant choices. The same points could be made about many other technologies not treated in this sampling: space technologies, communication technologies, transportation technologies, agricultural technologies, educational technologies . . . the list is endless. But none of these poses the new dilemmas of responsibility more vividly than the biological technologies on which we end this chapter. Artful intervention in the creation of new forms of life and the capacity in principle to engineer irreversible alterations in the human species itself, are symbols of the hitherto god-like powers to which implemented theoretical intelligence has attained. Such issues from ethics border on religion, our next topic.

CHAPTER 7

Technology and Religion

7.1. MYTHIC IMAGES OF TECHNOLOGY

Religions, whatever else they may be, are at least expressions of the most profound values of their adherents. Philosophers of technology cannot afford to overlook the molten core of value commitments from which, for many people, both ethics and fundamental beliefs derive. Since technologies are concrete embodiments of the state of knowing and valuing in any culture (1.8), religious commitments will have intense bearing on our subject.

That human technological prowess has long posed religious issues is shown in ancient Greek mythology. The story of Prometheus illustrates the ambivalence of archaic human sensibility in approaching the implementation of human intelligence with that most basic—and symbolic—of all techniques, the control of fire.

According to the story, human beings were deliberately left helpless by the gods, who determined to keep for themselves the boundless powers that go with the control of fire. Into this situation stepped the Titan, Prometheus, with sympathy for shivering human underlings. Deliberately flouting the will of the gods, Prometheus stole the forbidden fire from heaven and gave it to humankind, thereby starting the irreversible process that led to civilization. For his effrontery, Prometheus was captured by the angry gods, chained to a rock, and condemned to daily torture by a fierce bird who would tear and eat his ever-regenerating vitals.

Technology, seen and felt from this mythic perspective, is something wonderful but forbidden, transforming *Homo sapiens* into *Homo faber* and giving our species a unique position in the universe, arming us with the means to challenge the gods themselves. Prometheus, for his defiant daring, became a culture hero for the Greeks,

who forgave his *hubris*, or overweening pride, even as they shuddered at his torments. The torments themselves were not unwarranted. They were seen as the
predictable *nemesis* that follows any upsetting of balance by *hubris*.

The divine fire itself, now warming the hearths of civilized existence, has the
same two-sided character: it can fend off the beasts, give comfort and light—but it
can ravage and destroy its users equally well. The ancients from whose poetic
musings this myth originated could have had no inkling of how far the fires from
heaven might spread in future millennia, or how many of the powers once associated with the gods might be "stolen" by the Promethean spirit reincarnated in
modern techniques. Some today continue to admire the pride, daring, and fore-
thought of the Titan,[1] while others look with awe and fear at the *nemesis* he
brought upon himself; but, either way, there is no denying that the Prometheus
story strikes a deep chord of spiritual recognition for contemporary consciousness.
These images purport to show how things stand with the human race as regards the
decisive forces of the universe. They raise the finally religious questions for our
topic. Are our technologies ours by right or by theft? Are our god-like powers
evidences of defiance against our proper, humble place in the universe? Should we
rejoice in our implemented intelligence or be contrite? Ultimate values are at stake.
What should we worship? What in the end is the meaning of the human venture?

Other myths from ancient Greece suggest that human beings should be cautious
about overstepping human bounds. The story of Daedalus, the inventor, who, with
his son, Icarus, escaped imprisonment by flying away on wings fashioned from
feathers and wax, is bittersweet. Daedalus himself was saved by his aviation technology, but Icarus pushed his limits too far, flying too high and too close to the
mighty sun, where the *nemesis* of melted wax and a crash into the sea requited his
youthful *hubris*. Technology as such is not here condemned, but incautious or
excessive use is warned against. Another warning, given through the story of Midas,
who wished unwisely that he had the power to turn all things into gold—the symbol
of ultimate material security—is not so much directly against the abuse of technology as against the obsessive quest for the means rather than the ends of life.
King Midas was punished for his *hubris* by having his wish come true. Everything he
touched turned to gold, bringing the *nemesis* that all the things he should most
have valued—the taste of food, the texture and scent of flowers, the kiss of his own
beautiful daughter—were lost to him in gleaming but unresponsive metal.

The catalogue of Greek myths could be extended, but the point is clear: the
spiritual sensibilities of this great source of modern Western civilization were deeply
engaged with our range of issues in their own way. These mythic images are various
but the preachments are similar, and they resonate from human depth to human
depth across the ages. The awesome wonders of civilized arts and methods are
simultaneously desired and feared, and religious terror is shown at upsetting the
natural balances of the universe. Greed, heedlessness, excess are acknowledged as

---

[1]Garrett Hardin, *Promethean Ethics: Living with Death, Competition, and Triage* (Seattle,
Wash.: University of Washington Press, 1980).

part of the world's condition; so also are fitting punishments, brought about by natural and supernatural forces, which implacably maintain the physical and moral equilibrium.

7.2. JUDEO-CHRISTIAN EMBRACE OF TECHNOLOGY

The other great source of Western civilization, which mingled with the classical source by the fourth century, A.D., is the Judeo-Christian heritage. For us living in the modern Western world, the main religious issues of technology are those related to the biblical stories of Judaism and Christianity, considered here not for their literal content but for the larger symbolic meaning they may convey.

a. Technopolitan Liberation

One interpretation of the Judeo-Christian approach to technology, specifically applicable to modern science-led technology as well as to all earlier forms of implemented human intelligence, stresses the favor with which the biblical stories of origin view the human rule over nature. The book that offers the most forthright version of a theological affirmation of technology is *The Secular City* (1965, revised 1966) by Harvey Cox. Since Cox has modified his views several times since the initial publication of the book (in its revised edition Cox disclaims for it "scholarly balance or universal thoroughness"[2]), it would not be fair to identify the "Cox" of the following exposition with the full Cox, but such an exposition is needed if we are to appreciate the range of religious responses that technology has provoked within the Judeo-Christian tradition.

Cox begins with the invented concept of the "technopolis." It is his word for the actual condition, but also the trend or direction, of modern civilization:

Although the term is an artificial one, it reminds us that the contemporary secular metropolis was not possible before modern technology. Modern Rome and modern London are *more than* larger versions of their Augustinian or Chaucerian forebears. There comes a point at which quantitative development releases qualitative change, and that point was reached in urban development only after the modern Western scientific revolution. Manhattan is inconceivable without structural steel and the electric elevator. Technopolis represents a new species of human community. The fact that it is a neologism will remind us that it is not yet fully realized.[3]

Living in such an urban center is the distinctive experience of the modern world. It is "the place of human control, of rational planning, of bureaucratic organiza-

---

[2]Harvey Cox, *The Secular City: Secularization and Urbanization in Theological Perspective* (New York: Macmillan, 1965, rev. ed., 1966), p. xi. Copyright © 1965, 1966 by Harvey Cox. Reprinted with permission of Macmillan Publishing Company.

[3]*Ibid.*, p. 5.

tion—and the urban center is not just in Washington, London, New York, and Peking. It is everywhere."[4]

What makes the technopolis a compelling theological topic is, according to Cox, that it was made possible by and is the direct fruit of the radical Judeo-Christian revolution against imprisoning nature religions of ancient days. In those days, all nature was haunted by gods and spirits that restricted human freedom and inhibited scientific curiosity. Nature, gods, and the human race were symbolized as locked together in a closed system cycling endlessly and unprogressively together. But this was to end with the Hebrew story of the universe, in which the "enchantment" of nature is shown as finally broken. "The Hebrew view of Creation signals a marked departure from this closed circle. It separates nature from God and distinguishes man from nature. This is the beginning of the disenchantment process."[5] By such disenchantment, a new world view opened. Nature could at last be seen as offering utilities instead of divinities.

Whereas in the Babylonian accounts, the sun, moon, and stars are semidivine beings, partaking of the divinity of the gods themselves, their religious status is totally rejected by the Hebrews. In Genesis, the sun and moon became creations of Yahweh, hung in the sky to light the world for man; they are neither gods nor semidivine beings.[6]

Similarly, the animals and nature in general are cleansed of the inhibiting "kinship ties" of the older world view. "Just after his creation man is given the crucial responsibility of naming the animals. He is their master and commander. It is his task to subdue the earth. Nature is neither his brother nor his god. As such it offers him no salvation."[7] Without such a freeing spirit, science as we know it could never have arisen.

Science is basically a point of view. However highly developed a culture's powers of observation, however refined its equipment for measuring, no real scientific breakthrough is possible until man can face the natural world unafraid. Wherever nature is perceived as an extension of himself or his group, or as the embodiment of the divine, science as we know it is precluded.[8]

Thus, as Cox presents the case, biblical religion provided the vital spiritual preconditions for science and modern technology, and thus for the technopolis in which we live today.

The Judeo-Christian tradition can rejoice in its offspring. For the first time in history, there is real spiritual freedom possible, thanks to life in the technopolis.

[4]*Ibid.*, p. 4.
[5]*Ibid.*, pp. 19–20.
[6]*Ibid.*, p. 20.
[7]*Ibid.*, p. 20.
[8]*Ibid.*, p. 21.

First, we enjoy the freedoms of *anonymity* today that village and town life denies. "Urban man's deliverance from enforced conventions requires that he choose for himself. His being anonymous to most people permits him to have a face and a name for others."[9] Second, we have the privileges of *mobility* that open our minds to varieties in human possibility and keep us from settling lethargically into the status quo. Cox recalls the famous song, "How Ya Gonna' Keep 'Em Down on the Farm Now That They've Seen Paree?" "Those who have been drawn into the tradition-demolishing orbit of urban life are never quite the same again. They will always know that things *could* be different; they will never again accept the farm as given; and this is the seedbed of revolution."[10] Third, we are delivered from the spiritual sin of worshipping the idol of our own values, not merely by the mobility but also by the *pragmatism* of technopolis. Our beliefs are relativized. We are forced to recognize change, movement, newness—and our standard of truth for our new religious constructions will be how well and dependably they function for the spiritual uses to which we put them. The key thing to defend against is not pragmatism itself, nor the technical approach to issues, but is the over-narrowing of what counts as "useful."

The Gospel does not call man to return to a previous stage of his development. It does not summon man back to dependency, awe, and religiousness. Rather it is a call to imaginative urbanity and mature secularity. It is not a call to man to abandon his interest in the problems of this world, but an invitation to accept the full weight of this world's problems as the gift of its Maker. It is a call to be a man of this technical age, with all that means, seeking to make it a human habitation for all who live within it.[11]

b. A Burden of Guilt

The historian Lynn White, Jr. in an independent discussion,[12] supports much of Cox's analysis of the essential role of biblical religion in making science and technology possible. First, he agrees that the modern world as we know it is the result of the surprising, and comparatively recent, melding (4.3) in Western Europe of what we have called theoretical and practical intelligence. This was a startling and unique occurrence. "Science was traditionally aristocratic, speculative, intellectual in intent; technology was lower-class, empirical, action-oriented. The quite sudden fusion of these two toward the middle of the nineteenth century is surely related to the slightly prior and contemporary democratic revolutions which, by reducing

[9]*Ibid.*, p. 41.

[10]*Ibid.*, p. 46.

[11]*Ibid.*, pp. 72-3.

[12]Lynn White, Jr., "The Historical Roots of Our Ecologic Crisis," *Science*, CLV, March 10, 1967. Copyright © by the American Association for the Advancement of Science. This influential essay has been reprinted in many places. For convenience, citations here will be taken from Carl Mitcham and Robert Mackey, *Philosophy and Technology: Readings in the Philosophic Problems of Technology* (New York: The Free Press, 1972), pp. 259-65.

social barriers, tended to assert a functional unity of brain and hand."[13] The Western character of modern technology and science is strikingly clear. "Our technology has absorbed elements from all over the world, notably from China; yet everywhere today, whether in Japan or in Nigeria, successful technology is Western."[14]

The same clearly Occidental cast is present in the scientific thinking that is now so closely associated with this technology. Despite the existence of many other sciences in the past, "Today, around the globe, all significant science is Western in style and method, whatever the pigmentation or language of the scientists."[15] White finds it historically illuminating, therefore, to look into the characteristic Western background of practice and belief that led to this techno-scientific Occidentalizing of the modern world.

The principal context for the medieval West, from which all this sprang, was biblical religion. Like Cox, White finds the Judeo-Christian story of creation especially pertinent to the rise of the technologies (e.g., deep plowing) and scientific attitudes (e.g., detachment) that created the modern world.

Christianity inherited from Judaism not only a concept of time as nonrepetitive and linear but also a striking story of creation. By gradual stages a loving and all-powerful God had created light and darkness, the heavenly bodies, the earth, and all its plants, animals, birds, and fishes. Finally God had created Adam and, as an afterthought, Eve to keep man from being lonely. Man named all the animals, thus establishing his dominance over them. God planned all of this explicitly for man's benefit and rule: no item in the physical creation had any purpose save to serve man's purposes. And, although man's body is made of clay, he is not simply part of nature: he is made in God's image.[16]

Thanks to common belief in this story of creation, White concludes, Christianity had the effect in the medieval West of encouraging the technologies of extraction and exploitation that fitted well with the biblical view of human dominance. Everything in nature, he points out (as Cox also emphasizes), was intended as a utility for human use. This resulted in a thought-world in which all things revolve around human interests. "Especially in its Western form, Christianity is the most anthropocentric religion the world has seen."[17] The human drama not only dominates all history, the human species is set apart from nature by the divine gift of the "image" of God, the creator. The sacred groves of prior animistic religions now become fair game as firewood. "By destroying pagan animism, Christianity made it possible to exploit nature in a mood of indifference to the feelings of natural objects."[18]

[13]*Ibid.*, p. 260.
[14]*Ibid.*, p. 260.
[15]*Ibid.*, p. 261.
[16]*Ibid.*, pp. 262–63.
[17]*Ibid.*, p. 263.
[18]*Ibid.*, p. 263.

Thus far White and Cox are in basic agreement over the alliance between biblical religion and the techno-scientific world. From this point, however, they sharply diverge in their evaluations of the situation. Cox regrets but dismisses the environmental damage caused by modern technology as "essentially childish and . . . unquestionably a passing phase"[19] in our liberation from old tyrannies of the enchanted universe. White, however, strongly condemns the destruction of nature that biblical anthropocentrism has brought on the world. The crisis is deep and unprecedented.

We would seem to be headed toward conclusions unpalatable to many Christians. Since both *science* and *technology* are blessed words in our contemporary vocabulary, some may be happy at the notions, first, that viewed historically, modern science is an extrapolation of natural theology and, second, that modern technology is at least partly to be explained as an Occidental, voluntarist realization of the Christian dogma of man's transcendence of, and rightful mastery over, nature. But, as we now recognize, somewhat over a century ago science and technology, hitherto quite separate activities, joined to give mankind powers which, to judge by many of the ecologic effects, are out of control. If so, Christianity bears a huge burden of guilt.[20]

What shall be done? The answer to the guilt of Christianity is not to throw more technology at a religious problem; it is to reform our religion. As White insists, "What we do about ecology depends on our ideas of the man-nature relationship. More science and more technology are not going to get us out of the present ecologic crisis until we find a new religion, or rethink our old one."[21] Since foreign religions do not transplant easily, White suggests that we look to alternative forms of Christianity for antidotes to the excessive activism of Western versions of the faith. Greek Orthodoxy, for example, was not guilty of Roman domineering ways.

The key to the contrast may perhaps be found in a difference in the tonality of piety and thought which students of comparative theology find between the Greek and the Latin churches. The Greeks believed that sin was intellectual blindness, and that salvation was found in illumination, orthodoxy—that is, clear thinking. The Latins, on the other hand, felt that sin was moral evil, and that salvation was to be found in right conduct. Eastern theology has been intellectualist. Western theology has been voluntarist.[22]

But perhaps even closer to home, and therefore even more available spiritually to the dwellers in technopolis who see the need for religious reform, is help from St. Francis of Assisi. St. Francis, though himself a Westerner, preached by word and example against Western activism, materialism, and pride.

[19]Cox, *The Secular City*, p. 20.
[20]White, "The Historical Roots of Our Ecologic Crisis," p. 264.
[21]*Ibid.*, p. 264.
[22]*Ibid.*, p. 263.

The key to an understanding of Francis is his belief in the virtue of humility, not merely for the individual but for man as a species. Francis tried to depose man from his monarchy over creation and set up a democracy of all God's creatures. With him the ant is no longer simply a homily for the lazy, flames a sign of the thrust of the soul toward union with God; now they are Brother Ant and Sister Fire, praising the Creator in their own ways as Brother Man does his.[23]

With this appeal to a more gentle, modest, and contemplative religion, White concludes his analysis:

Both our present science and our present technology are so tinctured with orthodox Christian arrogance toward nature that no solution for our ecologic crisis can be expected from them alone. Since the roots of our trouble are so largely religious, the remedy must also be essentially religious, whether we call it that or not. We must rethink and refeel our nature and destiny.[24]

c. A Fuller Image of God

A Roman Catholic position that appeared before either Cox or White published their views, but one that could otherwise be taken as a response to White's appeal to rethinking and refeeling the Christian approach to technology from within the Latin tradition, places systematic emphasis on the concept of the "image of God" as normative for human destiny.[25] W. Norris Clarke writes of "technology as an element in the total development of man as an image of God."[26]

Much of religious and practical importance can be unpacked from this conception. First, the idea of the image should not, Clarke insists, be static. It is not to be worn like a badge of pride, but to be fulfilled in an evolving universe. The doctrine of the image claims its proper context in the story of creation, where it is the symbol of the right relation between matter and spirit. The creation story is vital, not so much for its mythic details as for its overall import:

Man's own origin and destiny now emerge not as a mere accident of landing on top of the heap of the world of matter by some lucky turn of the blind wheel of chance. They are the result rather of God's own creative activity, first bringing into being the material universe as a matrix and instrument for the development of the spirit of man, and then infusing each human soul into this evolving system at its appropriate time and place.[27]

---

[23]*Ibid.*, p. 265.

[24]*Ibid.*, p. 265.

[25]W. Norris Clarke, "Technology and Man: A Christian Vision," in *The Technological Order*, ed. Carl F. Stover (Detroit, Mich.: Wayne State University Press, 1963). By permission of Wayne State University Press. Citations here are for convenience taken from a slightly revised version reprinted in Mitcham and Mackey, *Philosophy and Technology*, pp. 247-58.

[26]*Ibid.*, p. 247.

[27]*Ibid.*, p. 250.

Given this dynamic emphasis, the doctrine of our being created in the "image and likeness of God" stands for our divinely given destiny to unfold and develop this image to the fullest possible extent in this life, in order to be united in eternal beatific union with him in the next."[28]

If this relationship between matter and the developing divine image is taken as primary for human destiny, then the technological phenomenon takes on a different hue. It plays a positive part in the drama of human fulfillment.

Man's self-development and self-expression through matter, with technology as his instrument, now appear not just as the satisfaction of some egoistic drive for power and self-affirmation, but as the fulfilling of a much higher and more sacred vocation, the God-given vocation to authentic self-realization as the image of God his Creator. The material world which is to be the object of man's technological domination is now seen not as some hostile or indifferent power that man has tamed by his own prowess and can exploit ruthlessly at his own will with no further responsibility save to himself. It is, rather, both a loving gift and a sacred trust to be used well as its giver intended, with a sense of responsibility and stewardship to be accounted for.[29]

More profoundly yet, the "image of God" metaphor itself begs for a fuller theological analysis. Who is this God, in the image of whom we humans are said to be created? What is He like? Is God simply a thinker or simply an active being? No, Clarke replies, God is in Himself a complete fusion of contemplation and action, and by direct analogy, so should we be, who are made to fulfill His image.

The notion of a dynamic image of God to be developed lends here a much deeper significance and dignity to man's cultivation both of science and technology. For man to imitate God, his Creator and—in the full Christian perspective—his Father, he must act like his Father, do what He is doing so far as he can. Now God is at once contemplative and active. He has not only thought up the material universe, with all its intricate network of laws, but he has actively brought it into existence and supports and guides its vast pulsating network of forces. God is both a thinker and a worker, so to speak. So, too, man should imitate God his Father by both thinking and working in the world.[30]

This means that we have a religious mission to exercise practical as well as theoretical intelligence. The full image of God cannot be fulfilled by the contemplative outlook alone. The human destiny is to work, "not to create some totally new world out of nothing, which only God can do, but to recreate the world that has been given him, malleable and plastic under his fingers, to be transformed by his own initiatives and artistic inventiveness, so that it will express in a new way both the divine image of its Creator and the human image of its recreator."[31]

[28]*Ibid.*, p. 250.
[29]*Ibid.*, p. 250.
[30]*Ibid.*, p. 250.
[31]*Ibid.*, p. 250.

What, though, of criticisms like White's, that the Christian encouragement to "recreate" this malleable world gives us a license to rape the natural order? Clarke strongly rejects any such accusation. Environmental heedlessness is no part of the Christian invitation to fulfill the image of God through matter. On the contrary, he says, "this notion of man as dynamic image of God, with the vocation to develop this image by an evolving dialogue with the material cosmos, sets technology in a wider framework which provides strong religious, moral, and humanistic controls on its exploitation."[32] True enough, Christians will be prepared by the doctrine of Original Sin to find distortions and dislocations in every part of life, including the technological, but this warning is not against the uses of technology, or the mission to work creatively with matter, but is against the *mis*uses that come from such spiritual deformations as "egotism, sensuality, self-indulgence, lust for power, wealth, pride, and self-aggrandizement."[33] The Christian will know that these can be anticipated quite apart from the technological phenomenon and will as a result not be naive about the dangers of technical *hubris* when it, too, appears along with all the other sins to which we are drawn. This gives to the theological vision a wise balance, since

the alert Christian, alive to the full implications of the Christian vision of man, will look on technology with a restrained and carefully qualified optimism, seeing it as at once a great potential good for man by nature and yet in the hands of fallen and selfish human nature an almost equally potent instrument for evil.[34]

Finally, the Christian doctrine of redemption, according to Clarke, brings out the "intrinsic goodness and dignity of matter itself,"[35] insofar as the incarnate God used matter through which to express His divinity. It is, further, significant that the God-man, Christ Jesus, was a carpenter. Divinity incarnate as a tool-user offers a perfect model for sanctioning the activity of practical intelligence shaping material to human purpose. "Thus the labor of the young Jesus as a carpenter in Nazareth already lends, in principle, a divine sanction to the whole technological activity of man through history."[36] True, the side of the redemption story that emphasizes the mortification of the flesh, as in Christ's crucifixion, seems to clash with technology's promise of material bounty and comfort, but, Clarke points out, the deeper purpose of mortification is "to establish in man the proper docility of matter to spirit."[37] Exactly this same purpose, controlling matter by mind, is fulfilled by a properly conceived technology.

Hence the spirit of Christian mortification can actually operate as a powerful con-

[32]*Ibid.*, p. 250.
[33]*Ibid.*, p. 251.
[34]*Ibid.*, p. 251.
[35]*Ibid.*, p. 251.
[36]*Ibid.*, p. 252.
[37]*Ibid.*, p. 252.

trolling factor for directing the use of our technological power along the proper lines for the authentic enrichment of man on the highest levels of his human capacities, instead of allowing it to be diverted toward the mere gratification of man's inferior appetites and desires and thus enslave him further to matter rather than liberate and elevate him above it.[38]

In the end this is the only finally adequate approach to the technological phenomenon, considering the power of our new tools and the persistence of sin. Clarke acknowledges the depth of the danger, but finally affirms the creative potency of technology under the guidance of God.

One might well say, in fact, that only men with something like the Christian virtue of self-denial, whether applied to sensuality or to egotism would really be safe enough to entrust with the responsibility of deciding in which directions to follow up the almost limitless potentialities made available to us by technology. The wider the range of possibilities open to a man's free choice, as we all know, the greater his need for self-discipline and selectivity, lest he destroy himself.[39]

7.3. JUDEO-CHRISTIAN REJECTION OF TECHNOLOGY

The three previous positions differed among themselves, but on one matter they were agreed: namely, that the Judeo-Christian tradition is (for better or for worse) basically supportive of the human technological enterprise. There are other interpretations of the biblical viewpoint, however, that see a sharply different picture.

a. Technology as Original Sin

One question that must interest any interpreter of the Genesis story of creation and of the subsequent stories in Old Testament literature is the attitude expressed there toward human powers of technical control. The interpretation[40] takes the view that at least one of the biblical story-tellers—the author known as J because of his preferred name for God ("YHWH" or "Jehovah") and because of his special emphasis on places and events in Judea—took a profoundly negative attitude toward urban, technological existence in general, preferring instead the simpler and purer values of nomadic life and stressing their ultimate significance.

In this interpretation, the fruit of the Tree of Knowledge of "Good and Evil," which God forbade to Adam and Eve, represents a sheep herding nomad's baleful view of the mysterious *practical* arts of sophisticated city dwellers. J deplored the "high technology" of that time. It was the sort of dangerous potency that should be left to God.

[38]*Ibid.*, p. 252.

[39]*Ibid.*, p. 252.

[40]This hypothesis was developed in lectures by the Old Testament scholar, J. Philip Hyatt, at the Vanderbilt Divinity School in 1954.

What else might the "knowledge of good and evil" have been? At the minimum any answer to this question should fit two clear criteria: first, this "knowledge" must be something properly characterizing God's own nature, and, second, it must be something that human beings ought not to have.

These two criteria, however, limit the field of possible answers more than is usually noticed. They mean, for example, that the fruit of the forbidden tree could not have been sexual knowledge, since that can hardly be taken as characterizing God's nature. And they mean that it cannot have been basic moral knowledge for two reasons: (1) it seems wrong to suppose that morality could have been bad for humans to possess, and (2) it is clear in the context of the story that God presupposes that Adam and Eve already had that kind of knowledge at the time of their temptation, insofar as they already knew that it would be wrong to eat what God had forbidden.

It is at least a plausible hypothesis, then, that the forbidden fruit was technological civilization itself. Further support appears as the story progresses. The first act attributed to Adam and Eve, for example, after eating this fruit, was to sew themselves garments, inaugurating the clothing industry. Their murderous son, Cain, is described as founding the first cities.[41] Then the people of the cities of Cain used their knowledge to build the heaven-challenging Tower of Babel, a clear example of technology so advanced as to worry God lest the human race become too potent.[42] Jealous of such knowledge, God confuses the languages of the earth to prevent human unity and thus to undermine—until recent times—cooperative technologies on this scale.

The one approved technology in the early Old Testament is Noah's Ark, for which the detailed construction plans were prepared by God, not Noah.[43] The Ark itself was simply to provide for a saving remnant of the original creation while human pride and technological urban power were being scrubbed from the earth by the Flood.

Later, at the time of King David, God is shown as angry at David's plan to take a census.[44] The overweening knowledge and powers of social control this could give human rulers is seen, on this hypothesis, as yet another presumptuous bite from the forbidden fruit.

b. Technique as "Fallen"

A closely allied interpretation of the biblical view of technology is urged by the eminent French social theorist and theologian, Jacques Ellul, who strongly condemns efforts to sanctify technology on biblical grounds. These grounds, as we have observed above, sometimes play on the idea that the original commission given to

[41]Genesis 4:17.
[42]Genesis 11:1–9.
[43]Genesis 6:14–16.
[44]II Samuel 24:1–17.

Adam before the Fall was to have legitimate "dominion" over nature. This much is granted. Ellul, however, makes a sharp distinction between the sort of dominion intended by God before the Fall and the sort of dominion by technique that we have exercised since the expulsion of humanity from Paradise.

Is technology then necessarily "fallen"? Some may argue to the contrary, claiming that technology was symbolically blessed by being required even in Paradise. They quote the verse saying that Adam was put into the Garden of Eden "to till it and keep it" (Genesis 2:15, RSV). Then they ask how Adam could have "tilled" a garden without the tools of agricultural technique.

Ellul, however, in harmony with Calvin and the Reformed tradition, stresses the depth of the "break" between the blessed sinless situation in the Garden of Eden and the actual corrupt human condition within history. Our minds, on the wrong side of the break, are too corrupted by sin to know what paradisiacal existence was like for Adam and Eve. But we can at least be confident, Ellul holds, that they did not need tools to do their tilling and "keeping" or guarding. In Paradise, he points out, it is inconceivable that weapons would be needed for the latter. Why, then, should tools be needed for the former?

If Adam needed tools for cultivation, then he also needed weapons for guarding. The two things are identical. If Adam's work was the point of departure, the beginning and the justification of technique, then his mission to guard was the point of departure, the beginning and the justification for police and armies. Is this not unlikely? It could not be more so. And if we reject weapons then we have to reject tools as well.[45]

Hard though it is for us to imagine, Adam before the Fall would have been able to accomplish everything without effort. "No cultivation was necessary, no care to add, no grafting, no labor, no anxiety. Creation spontaneously gave man what he needed, according to the order of God who had said, 'I give you . . .' (Genesis 1:29)."[46]

To think otherwise, Ellul argues, is to demean the perfection of creation itself, before the radical intrusion of sin. If anything is clear, he says, it is that "creation as God made it, as it left his hands, was *perfect and finished*."[47] It is not the case that God left things undone for human creatures to complete. That idea of the human mission as "co-creator" may be innocently intended to do honor to our species, but "in this good intention there is always an honor stolen from God."[48] We must humbly realize that before the Fall, "God's work was accomplished, that it was complete, that there was nothing to add."[49] Therefore technology, when it

[45]Jacques Ellul, "Technique and the Opening Chapters of Genesis," in *Theology and Technology: Essays in Christian Analysis and Exegesis*, eds. Carl Mitcham and Jim Grote (Lanham, Md.: University Press of America, Inc., 1984), p. 129.

[46]*Ibid.*, p. 126.

[47]*Ibid.*, p. 125.

[48]*Ibid.*, p. 126.

[49]*Ibid.*, p. 125.

appears, is always a sign of the realm of necessity rather than love. It is always and only present in the condition of sin.

Such is the case for science too. Some may claim to see in God's allowing Adam to name the animals some early preparation for science, which classifies living forms and assigns them names. But Ellul rejects this notion hotly:

It is this kind of slipperiness that radically falsifies the meaning of a text of revelation. As for saying that the act of naming is the origin of science, this is a dramatic misunderstanding of what Adam did. . . . The language used by science is a language of division. Yet biblically, to give a name is exactly the inverse phenomenon: it is a re-capitulating fact. To assign a name is to discern a spiritual reality. . . . Biblically, we have a spiritual act which has nothing in common or no point of contact with the intellectual operation of science.[50]

Domination, not love, is the essential mood of science and technology, Ellul holds. That is why the combination of the two in the "modern project" gives no foothold for Christian admiration or approval. It is essentially and necessarily flawed.

The modern project has been different; *therefore*, the very reality of its working out in methods and forms of knowledge has been different. We cannot glorify God though the splitting of the atom, not through the manufacture of new chemical products, etc. Our experience could not have been other than what we made it. The exploration of the world in the 15th century, for example, directed by the will to power and exploitation, *could not* have produced any result other than the one that has been lived (colonialism). That is to say, it could not have been directed toward the peaceful admiration of creation and the adoration of the creator. This was not a live option.[51]

The illegitimate fruits of the modern project have perversely ripened in human pain and ecological destruction. Sin is indivisible. God loves creation, "with all that it includes, its variety, its blossoming; and it means that the ecological destruction is of the order of sin—as considerable as war, genocide, the exploitation of man by man, injustice."[52] What Christians and Jews—the people of the biblical revelation—could and should have done in this modern period was to have taken "a strict position on the limits of scientific pride, the application of techniques and the exploitation of nature."[53] Instead, apologists for technology within the churches deflected this mission of "holiness" by attempting to justify it religiously. Followers of God's revelation had a special revelation and hence a special responsibility.

[50]*Ibid.*, p. 130.

[51]Jacques Ellul, "The Relationship Between Man and Creation in the Bible," in Mitcham and Grote, *Theology and Technology*, pp. 142–43.

[52]*Ibid.*, p. 152.

[53]*Ibid.*, p. 153.

Instead, from the moment of the splitting of the atom, when certain Christians wanted to raise the question whether it was not exceeding the order God gave to his creation, whether it was not to penetrate into the secret of God, the Churches have replied that science had the right legitimately to do anything. Of course! Thirty years later we are threatened with extermination, more or less shortly, through the development of nuclear plants.[54]

7.4. JUDEO-CHRISTIAN CAUTION TOWARD TECHNOLOGY

Tones of praise and denunciation are not the only ones theologians can use when speaking about technology. Even a generally positive voice like Norris Clarke's (7.2.c), as we noticed, includes warning modulations. There are other Christian interpretations of technology that make it a prime goal to maintain the delicate balance between blessings and curses.

Egbert Schuurman, A Dutch thinker who has also engaged, as a senator, in the political life of his country, works theologically to achieve just such an equilibrium. Schuurman shares a Calvinist heritage with Ellul, but parts company with him and his followers at the point where Ellul's opposition to modern technology seems to tilt toward despair. Schuurman notes:

Ellul and his followers tend to assert that the problems of modern technology are too heavy to bear. As Christians they think that modern technology is an autonomous, demonic power. In *The Technological Society*—but also in later books such as *The Ethics of Freedom* and *The New Demons*—Ellul seems to argue that man is not the master of technology, but its slave and its victim. Man is the victim of a universal, artificial, monistic, self-directing power.[55]

This despair, Schuurman objects, is non-biblical, since it "does not leave open the possibility of deliverance."[56] Seeing clearly how difficult it is to control modern technology, Ellul has nevertheless exaggerated its autonomous powers. Schuurman acknowledges that computer technologies, for example, are now immensely difficult to control. But this difficulty represents only *relative* autonomy for technology in general.

It must be created and directed toward some use by someone. Technology is not an autonomous force. The fact that the development and direction of technology proper are guided and set by norms given to it from the outside, and the fact that it is precisely human beings who do this, make it clear that technology is a dependent phenomenon. Nevertheless the problems are great and heavy to bear.[57]

[54]*Ibid.*, pp. 153–54.

[55]Egbert Schuurman, "A Christian Philosophical Perspective on Technology," in Mitcham and Grote, *Theology and Technology*, pp. 107–08. By permission of the University Press of America, Inc.

[56]*Ibid.*, p. 108.

[57]*Ibid.*, p. 108.

In selecting only dark colors for his portrait of modern technique, Ellul has painted a picture that lacks the brighter biblical hues of promise. Schuurman offers an alternative view of history from a biblical perspective in which he hopes the whole spectrum will appear. This view has four major elements, presented in two pairs of principles that collide with and balance one another. The first balancing pair are, on the one hand, (1) *the cultural mandate* from Genesis—"Man has received the calling to dress or to build the creation, and to keep it (Genesis 2:15)"— and, on the other hand, (2) *the fall*—"Man forsook his original task"—after which the cultural mandate was no longer possible for human efforts, history ceased to be wholesome, nature became threatening, and human technology was distorted by sin.[58] "Skills and techniques of all kinds may be admirable, but the tyrannical or greedy use of human power over nature is a failure deriving from human sin, not from God's intention in the creation."[59]

The second balancing pair, however, are (3) *redemption*—"God himself provides redemption" and breaks the power of sin over history decisively through the coming of Jesus Christ—and, at the same time, (4) *disobedience and secularization*— "not everyone lives within the dynamic power of the creation, which is in Christ the power of the Kingdom of God."[60] This last reality, though serious as representing the continuation of the effects of the Fall even after the decisive victory over sin by Christ, is not properly separated from the context of the good news of redemption. Ellul runs the danger, Schuurman warns, of taking (2) and (4) without the good news of (1) and (3). The dangers of sinful disobedience have indeed been greatly reinforced by the powers of modern techno-science, but despite its intimidating face, contemporary technological secularization cannot finally force the meaning of history against the will of God.

It is a constant consolation to know that man on his own and by himself cannot make the meaning of creation, the Kingdom of God, impossible. On the contrary, the fact that the Kingdom of God is already on the way means that at any moment people may be converted and led once again to seek the Kingdom—even in a technological society.[61]

For Schuurman this divinely guaranteed possibility of conversion is the key to a balanced Christian doctrine of technology. Human motives have led to current techniques. To the extent that our motives have been dominantly anthropocentric and exploitative, our technology has been and will be a curse; to the extent, on the other hand, that new motives replace the old ones, our technology could be a blessing.

It is vital that Christians keep alive the real possibility of the human race returning to the God-centered motives symbolized in the original responsibilities of

[58]*Ibid.*, pp. 109–10.
[59]*Ibid.*, p. 110.
[60]*Ibid.*, pp. 110–11.
[61]*Ibid.*, p. 111.

Adam: to "dress" and to "keep" the creation (Genesis 2: 15). "Dressing" ("tilling") the creation, first, is an active motive, reflecting the Principle of Beneficence (6.1.c) by seeking to increase good. "Keeping" ("guarding") creation, second, is a preservationist motive, reflecting the Principle of Non-maleficence (6.1.c) by seeking to avoid doing harm.

Scientific technologies powered by such motives can serve God, humanity, and nature. There is nothing intrinsically wrong with science, Schuurman notes. Wrong motives, the lust for human power and autonomy, have corrupted something that could be a great good. Right motives could change everything for the better. "Guided by the right motive, man in his cultural activity can be a blessing for nature (1Kings 4:33-34) and at the same time enter into an open way toward the future."[62]

Religious persons who wish for a patron saint for this middle-of-the-road approach to technology will probably not be satisfied with Lynn White's proposal of St. Francis of Assisi (7.2.b). St. Francis, as Ian Barbour points out, stands for a beautiful spirit of unity with nature. That may be a good symbol for the human preservationist attitude, but it does not fit the inevitable manipulation that goes with the "dressing" or "tilling" on the active side of the religiously motivated technological life. In place of St. Francis, the equilibrium symbolized by St. Benedict might be sought. Barbour, who like Schuurman looks carefully to avoid extremes in theological attitudes toward technology, comments:

Compared with St. Francis's deep feeling and sense of union with the natural world, St. Benedict's response was more practical, using nature with care and respect. The Benedictine monasteries combined work and contemplation. They developed sound agricultural practices, such as crop rotation and care for the soil, and they drained swamps and husbanded timber all over Europe. Benedictines were creative in practical technologies related to nature.[63]

Debates among Judeo-Christian thinkers on the religious meaning of technology will continue to reverberate. Despite a shared set of biblical images and common acknowledgement of their ultimate importance, different principles of biblical interpretation are in use, different standards of authority, different assessments of the compatibility of religion and culture. Religious beliefs are rooted in one's most comprehensive and intense evaluations. Differences in evaluative emphasis can result in significant divergences in doctrinal position, just as differing doctrinal nuances can reinforce conflicts in value priorities. For philosophers of technology, seeing our topic bathed in the intense light of such ultimate commitments can be clarifying of the many values in this complex phenomenon, all of which beg for recognition and ordering. One's own values, too, through the discipline of dialogue

---

[62]*Ibid.*, p. 117.

[63]Ian G. Barbour, *Technology, Environment, and Human Values* (New York: Praeger, 1980), p. 25.

with such rival positions—finding oneself approving here, rejecting there, and allowing oneself to ponder these intuitions—may also be thrown into ever sharper perspective.

7.5. NON-WESTERN RELIGIONS AND TECHNOLOGY

A challenge to Western sensibility itself is offered, finally, by comparison with ultimate value intuitions from other cultures. Lynn White is probably correct in thinking that the effective traditions in any ongoing civilization will be those that are internal to its history (7.2.b), thus capable of striking responsive chords. Reform and revival movements are not unknown. Schuurman's anticipation of widespread conversion, though theologically based, is not entirely without historical precedent.

It is not likely, therefore, that large numbers in Europe and America will adopt Taoism or Buddhism or Hinduism, although they are major living religions in our time. It is still less likely (despite the ecological attractiveness of shamanism and totemism to some contemporary commentators[64]) that many will convert to the nearly extinct religions of American Indian tribes. It is useful, nonetheless, to be aware of radical alternatives to the religious and metaphysical assumptions that Westerners tend to make. As we heard from Whitehead at the outset, we observe by the method of difference (1.1), and if we are to be clear about our own fundamental attitudes and beliefs, it helps—at a minimum—to know in what way they differ from other possibilities. Beyond this, who can tell what influences will "take" in the twists and pressures of historical events yet to come? It may be that some day a selection from or combination of the following views, or others, may become our own.

a. The Tool and the Tao

A profound intuition in one strand of Chinese thought, Taoism, declares the underlying unity of all reality and the fluidity of all forces. If nature is pushed at one point, there will be a counter-push at some other. The "Yin" aspects of reality must be in equilibrium with the "Yang." The best action may be inaction.

Such a supreme valuation of "letting things be" is in obvious contrast to the active stance even of Westerners who think of themselves as self-restrained in their stewardship of nature. There are Western contemplatives who come close to the studied passivity of the Taoist. One voice that blends the Oriental with the Occidental is P. Hans Sun, a Catholic monastic born in China between the worlds of Christian missionary and Chinese diplomat. Sun brings the Taoist principle of nonaction into his Christian theological critique of technology by identifying prayer

---

[64]Theodore Roszak, *Where the Wasteland Ends: Politics and Transcendence in Postindustrial Society*, Garden City, N.Y.: Doubleday, 1973), especially Chapter 4.

with non-action and technology with action. He opts for prayer. "The opposition between theology and technology is thus at its foundations an opposition between non-action and action. In light of the fact that ultimately non-action will prevail over action, it seems at least reasonable to suggest that choosing the latter runs the danger of succumbing to bad faith."[65]

On the scale between aggressive Beneficence and passive Non-maleficence, Taoism leans unreservedly toward the latter. In contrast, the Christian who thinks that it is a part of the religious responsibility of humanity to "rule" creation, even to improve it, is required to reflect on the basis of this belief. Does Western religion tend to put "doing good" ahead of "refraining from doing harm"? What is more important? What are the strengths and possible weaknesses of each ultimate valuation?

b. Buddhism and Detachment

Another great religious perspective, the "Middle Way" of Buddhism, challenges the importance that Westerners, including even mild-mannered Christian environmentalists, tend to place on their doings and abstainings. The Eightfold Path taught by the Buddha has a place in it for practical life. This religion, in its main forms, refuses to turn its back on the world in any extreme of asceticism. But while in the world, the Buddhist is not to be attached in mind or heart to the world. The greatest goal of all is the capacity to be unconcerned, to be spared from suffering the pains of caring and striving and not to suffer even from concern about the goal of striving after . . . the goal of not striving!

There could hardly be a less "technological" attitude than the Buddhist way of non-attachment to what is done or to ways of doing. Technology as the implemented practical expression of intelligence is not condemned, but neither is it sought. "Right thinking" on the Buddhist path is far removed from the problem-solving eagerness of the Reason of Ulysses (3.3), and almost equally remote from the thirst for detailed theoretical coherence and adequacy of the Reason of Plato (4.1). The entire techno-scientific enterprise comes, from a Buddhist perspective, not under criticism or even scorn—those are too-much-caring Western concepts—but, instead, under the more radical (and difficult) challenge of irrelevance. What is it in practical life, or life itself, that is important enough to warrant the sufferings of attachment? All Western religions seem to put a positive value on personal and social existence—"to live, to live well, to live better," under some interpretation (spiritual or secular) of the terms—but is this a mere cultural prejudice? The Buddhist perspective forces Westerners back to reconsider even such basics of Western thought.

[65]P. Hans Sun, "Notes on How to Begin to Think about Technology in a Theological Way," in Mitcham and Grote, *Theology and Technology*, p. 191.

c. Hinduism and Illusion

Other millions of persons find in Hinduism a faith for organizing life and giving shape to thought about the ultimate character of things. There are many varieties of Hinduism (of which Buddhism, at the time of its origin, was one), but beneath the protean flux of Hindu imagery and ritual there is a basic conviction that the character of things around us is unstable and fundamentally unreal. The visible world is *maya*: unreliable, tricky (from this Sanskrit root the English word "magic" descends).

Technology, then, as technique embodied in the unreal world of *maya*, can only deal with the surface of things. What is important, however, is not this surface variety but the underlying and unifying absolute, Atman, the All, which is beyond time and change and beyond scientific methods of knowing. Science studies mathematical forms of what is basically unreal; technology tinkers with appearances. From a Hindu perspective, the whole Western enterprise, the techno-scientific project itself, is a mistake, rooted in illusion. What makes us think the world of nature is so real?

This is a curious question—for metaphysics. Thus it leads directly into our final chapter. Several metaphysical issues have arisen as corollaries of religious commitments in this chapter, directly or indirectly. The *reality of the world of nature*, thus the epistemological reliability of science and the firmness of technology's grip on what matters, is one of these questions. The *nature of human nature* is another. The *reality of human freedom* in the face of apparent technological determinism is a third.

Religion is not identical to metaphysics, since religion as an expression of worship is fundamentally a question of values. But beliefs or theories about ultimate reality are often implicit and influential in debates over ultimate values. And the situation is reciprocal: ultimate value intuitions also often influence theories about the basic nature of reality. The topics of these chapters are "final" in more than one sense. They involve one another. They involve those who ask them. There is no getting beyond them. Neither is more fundamental than the other, any more than axiology is more fundamental than metaphysics, or metaphysics more fundamental than axiology (1.3.b and 1.3.c). The metaphysical questions, however, have yet to be faced explicitly in this book. To them we may now finally turn.

CHAPTER 8

Technology and Metaphysics

8.1. DOES MATTER MATTER?

It might appear—at least at first—that a great deal hangs on the metaphysical question with which the last chapter ended: Is the visible and tangible world real or illusory? We are asking, after all, about the material world that technology confronts, controls, and changes. More, since the rise of distinctively modern technology, we are also asking about the reality of the subject matter of science. If we should become convinced, like the Hindus (7.5.c), that the world of daily life is mere appearance, *maya*, what would become of science and technology? What would be the implications for the entire edifice of the modern world?

The first task of a philosopher, before trying to answer such a "global" question, is to get clear on what, more precisely, is being asked. What is meant by "mere appearance"? What is the force of "illusion"? How should we understand "reality"?

We might start by recognizing that there is a biologically important sense in which the question of appearance versus reality can be raised. It is the sense, rooted in our evolutionary heritage, in which it is vital for practical intelligence to distinguish between how something seems and how it really is. Nature is full of disguises, camouflages, and traps. This appears to be a twig; but is it really a worm I could eat—or a praying mantis waiting to eat me? The meadow appears safe; but does it really hide a lion in the tall grass? Beyond those dunes appears to be water; but is it really a mirage? Adaptive advantage came to those early forebears of our species who paused to test strong-appearing vines to see if they were really strong enough; successful spear fishing depends on compensating for the difference between where a fish appears and where it really is. The history of social life has taught the human race to beware of Greeks bearing gifts and of disciples giving kisses. Things are not always what they seem.

a. A Naïve Question

Could this be the sense in which metaphysical alarm is raised by asking whether the material world as a whole is no more than appearance? If so, there should be discernible consequences for those unwary enough to treat appearances as though they were reality. In all other cases this is how we learn the difference: the vine breaks, our spear misses the fish, the war is lost, we are betrayed. In all such cases what we mean by something's being "mere appearance" is defined by its *unreliability in supporting expectations about the future.* From some part of our experiences we come to anticipate certain other experiences but we have a different set of experiences instead. Whatever is *reality* in this sense is defined by the final set of experiences that provide the "bottom line" for the question. What turns out to be *appearance* is the set of experiences that somehow led us astray.

Our comprehensive metaphysical question about the "unreality" of the entire world of experience, however, cannot possibly have such a meaning. The set of experiences by which we would discover that all our experiences were unreliable would themselves be part of our experience, and thus (by hypothesis) would be unreliable. Without some reliability to provide a "bottom line" to ground the distinction, there is no sense in which unreliability can be defined.

The metaphysical issue, then, cannot simply be a question of whether nature will have surprises for us in the future. Scientists alert for such anomalies can use them to force revisions in current theories, to amend current laws, or even (reluctantly) to conclude that some events may be irreducibly statistical, or matters of chance. The question of the adequacy of our current predictions is an empirical question that is open in principle to further careful, open-minded observations. But no observations could even conceivably shed light on the question of whether all observations are merely "appearance."

The philosopher, George Berkeley (1685-1753), pushed these logical points to their fullest in his attack on the "prejudice" that somehow belief in a material world is required to account for science and ordinary experience.[1] Since all our confirming evidence is ultimately experiential—i.e., lies in the realm of ideas—then what meaning can be given to something like matter, something supposed to lie "behind" or "outside" the realm of actual or possible experience? Matter literally does not matter. Since our ideas are all we can directly experience, the whole meaning of reality can and must be expressed in experiencing or being experienced. When Dr. Samuel Johnson attempted to "refute" Berkeley by kicking a stone,[2] all he succeeded in confirming was that he had certain predictable perceptions in his

[1]*Berkeley's Philosophical Writings*, ed. David M. Armstrong (New York: Collier Books, 1965).

[2]Frederick Copleston, S.J., *A History of Philosophy* (Garden City, N.Y., Image Books, 1964), Vol. V, Part II, p. 12.

foot and leg, along with perceptions of sound and sight cohering with one another. He could not prove thereby that in addition a world of unexperienced matter was "real."

Since this is so, what difference does it make either to claim or to deny that the material world, the world of implements and artifacts, is "real"? Technology will not be any more reliable if the world is "real," or any less reliable if it is "unreal." Scientists would not observe anything different if it is "real" than they would if it is "mere appearance." There is nothing more to observe than the entire observable world. The claim that the world of science and technology is or is not "ultimately" illusion thus turns out to be empirically empty: Neither answer to our question can in principle offer us anything different to expect. But if so, then the question itself—naively put—is not logically a normal question at all, since it is not asking anything that could have a normal answer.

b. A Scornful Dismissal

There is little wonder, under these circumstances, that in this century of techno-scientific thinking, many philosophers, swinging to the opposite extreme, came to the conclusion that such questions—however put—ought not to be taken seriously by philosophers at all. Metaphysical questions became "controversial" (1.3.c) not so much for the different answers that they seem to solicit as for the logical status of the questions themselves. If they are empirically empty—if any "answer" must be unverifiable (in the sense of not offering any specific set of observations by which we could determine its truth) and/or unfalsifiable (in the sense of being compatible with any observations at all)—then asking them must be serving some other function than seeking normal knowledge. The Logical Positivist movement of the mid-twentieth century concluded that the actual function of metaphysical questioning is merely emotive—to express or elicit great global feelings.[3] Those who ask or try to answer such "questions" are at minimum confused by the superficial grammatical similarity of their vacuous utterances (e.g. "Is the world real?") to genuine factual questions (e.g. "Is the dinner ready?"). If they persist in such metaphysical talk after being shown these differences, some Logical Positivists suggested with a touch of glee, it must be because there are emotional needs—perhaps feelings of deep insecurity—that are being expressed thereby.[4]

c. Taking Stock

The Logical Positivist analysis of metaphysical issues has its point. Metaphysical questions do not function like "normal" factual queries. They are by nature so comprehensive that they defy the usual rules for asking or answering questions. Normally we want answers about things, like Whitehead's elephant (1.1), that might

---

[3] A.J. Ayer, *Language, Truth and Logic* (London: Victor Gollancz, 2nd ed., 1946), chap. I.
[4] Morris Lazerowitz, *The Structure of Metaphysics* (London: Routledge and Kegan Paul, 1955).

or might not be present. Metaphysical questions may not be amenable to the "method of difference."

Likewise, metaphysical questions often carry a strong valuational charge. If we say that something is "unreal," ordinary language, founded on ancient associations, carries an inevitably negative impact. What is real is important. It represents the "bottom line." It has the last laugh. What is unreal or "mere appearance" is also negatively important because it is dangerous when it conceals the real. It can cost a meal, a friend, a war. The power of the word "artificial" to damn something, as we saw (2.8.a), rests on the implicit value judgments echoing in the words "unreal," "false," or "fake."

Therefore we can grant that metaphysical questions are not "normal" ones, and we can agree that these questions carry powerful valuational weight. It is another question, however, whether the "method of difference" is the *only way* of thinking our way toward responsible answers for questions of great scope. And it is still another question whether our value judgments—though deeply felt—are *nothing more* than expressions of emotional feeling. What we need before proceeding any further is a richer theory of metaphysical thinking.

8.2. METAPHYSICAL THINKING

A helpful way of looking at metaphysical thinking is to check what features it has in common with other sorts of theory, and what special traits it may possess. What, at root, is a theory? Stripped to its essentials, a theory can be any "abstract calculus" or system of concepts, purely internally defined in terms of the symbols arbitrarily chosen for the system, with no references outside itself.[5] Such a symbol system would count as an "uninterpreted" theory, or a theory not yet *about* anything. These systems are of great interest to logicians and mathematicians, who concern themselves with ways of developing such formal calculi so that they will have proven characteristics of logical consistency and internal coherence—the qualities that are the essential internal requirements of all effective theorizing (1.4.a and 1.4.b). Scientists and others who want to theorize about some subject matter may adopt and use these abstract calculi as the structural skeletons for their thought, interpreting the symbols with concepts drawn from some area of interest. The interpreted theory then allows its users to draw conclusions, make comparisons, establish connections, and (sometimes) make predictions about the subject that is articulated by the theory. Once a theory is interpreted by a subject matter, it becomes important to make sure that the theory is as adequate as possible to its subject matter (1.4.c.).

[5]Ernst Nagel, *The Structure of Science* (New York: Harcourt, Brace and World, 1961), pp. 91-93.

a. Models in Metaphysical Thinking

Often theories are interpreted in terms of models. A model "interprets" a theory for a specific subject matter, while the interpreted theory, in return, "articulates" (literally, "provides joints for") its model by means of its formal structure. The model, as it were, offers its theory something to theorize about; the theory offers a crisp set of interrelated abstractions as a skeleton on which the model "fleshes out" a complicated or obscure subject.

Models come in . . . many models! The one thing they all do is to represent something else. We can distinguish them in terms of their *type*, their *scope*, and their *function*.

(1) *Type*. The first familiar type, made perhaps of plaster, clay, wire, wax, wood, etc., is the *physical* model, of which some may actually work to display processes; this working would distinguish them as *mechanical* models. The other great type is the *mental* model, which could range from remembered physical objects or events, i.e., *imaginary* models, to freely conceived subject matters (like "black holes") that could never be built or actually perceived, qualifying them as *conceptual* models. The most highly abstract conceptual models, dealing with numbers or geometrical properties, make up the so-called *formal* models.

(2) *Scope*. Models of various types can represent a *unique* referent (the Statue of Liberty), a *class* of things (the hydrogen atom), or even a vastly *comprehensive* subject matter (the whole physical universe). The scope of metaphysical models, as we shall see, is definitive and crucial for the logic of metaphysical thinking.

(3) *Function*. There is endless variety in the possible functions of models. These will vary with different purposes: e.g. using a model as a souvenir, relying on a model as a guide for copying, and the like. What interests us especially in this chapter is the function of models in theorizing.

What models do for theory also varies with the purposes of the theorizer, but to take scientific thinking as an example, there are three principal contributions. First, interpreting a theory with a model can provide an otherwise abstract theory with *conceptual definiteness*. If theories are meant to provide understanding for human thinkers, there must be limits to the theory's reliance on pure abstractions. Models offer intuitive grasp. Second, since a model is likely to be chosen from among things that are first encountered as unified entities, such a model can help to *suggest a pattern of conceptual connections* within the subject matter being modeled. One of the principal motives of theorizing is to understand a subject matter, whatever it may be, as a whole. Third, a scientific model can *suggest new areas of discovery*, since there may be interesting features of the model that have no corresponding formalism (yet) in the articulating theory. A fortunately chosen scientific model may therefore stimulate growth in the theory it interprets, leading to new empirical investigations and to the uncovering of hitherto unknown features of the world.

Metaphysical theorizing can now be mapped in terms of these general remarks about theories and models. First, it is a type of theorizing, and therefore shares the fundamental obligations of all theorizing to logical consistency, internal coherence, and experiential adequacy. Second, its models tend to be mental and normally conceptual in type.[6] Third, the distinguishing mark of metaphysical thinking is its unlimited comprehensiveness in scope. Beyond even the vast comprehensiveness, for example, of astronomical-cosmological theories and models—which deal with the entire physical universe, the processes of its origination, its geometry, its dynamics—metaphysical theories and models include all these in their scope and continue to ask whether there is anything more even than the physical universe.

This *unlimited comprehensiveness* of metaphysical theorizing is its defining mark, preventing it by its all-inclusiveness from relying on the "method of difference" and resulting in its principal distinctions from scientific thinking. All the sciences are, by self-definition, limited in their subject areas. The biological sciences do not attempt to include astronomical phenomena; and astronomy, despite its spatial all-inclusiveness, does not try to account for such areas as, for example, cell division. Even physics, by which many thinkers hope eventually to explain indirectly all processes and events (including cell division) that can be observed in the physical world, does not attempt to deal *as such* with the phenomena of life or psychology or society—and not at all, of course, with supposed non-physical entities of religion or mathematics. Any all-inclusive framework for the sciences will itself be beyond any of them. Philosophically oriented scientists, like Einstein, sometimes reflect thoughtfully on the overall significance of their work for metaphysics, and metaphysicians often study the sciences to learn what the best current theories and models dealing with parts of reality may have to teach them—but also to subject the abstractions of those theories to such philosophical critique as they may deem necessary.

The standards of metaphysical critique used will need to be appropriate to the comprehensiveness of the metaphysical enterprise. That means that consistency, coherence, and adequacy will be sought on a grand scale. Scientific abstractions that fail to cohere with a metaphysical conception of the whole will inevitably be held at arm's length, in anticipation that alternative theories or even scientific revolutions will change the picture; and scientific theories that depend upon the idealization or simplification of data considered important for philosophical adequacy will be taken to task.

Metaphysical construction, like metaphysical critique, aims for *systematic connection* rather than empirical discovery. Here we find a major contrast between the function of models in science and in metaphysics. In the latter, conceptual models can, because of their unlimited comprehensiveness, have two of the functions of "normal" scientific theorizing, but not the third. That is, a metaphysical model— one drawn from a significant portion of reality and applied to the whole of reality in an attempt to understand it—can offer a thinker conceptual definiteness for an

[6]Benedict Spinoza, who took his principal formal models from geometry, may count as an exception.

otherwise overwhelming subject and a pattern of conceptual connections to suggest for the subject the unity that all theorizing seeks. What it cannot provide, since everything that is or could be must already be accounted for in an adequate metaphysical theory, is the heuristic guide to future empirical discoveries or unanticipated predictions that scientific models can offer. The philosophical investigator, in other words, cannot hope to pick up a telescope, like Galileo (4.2.a), and make new observations that will confirm or confute a metaphysical world view. Galileo's opponents at first felt confounded in their metaphysics only because they had allowed their world models to become too intimately entangled in an eventually falsifiable scientific system.

All this means that metaphysical thinking is indeed different from the "normal" sorts of thinking in science or everyday life that can be verified or falsified by some specifiable set of observations. It is instead highly systematic, commending itself theoretically on the basis of how well it allows everything to hang together, to "make sense" in a large pattern that reflects and "rings true" to the most pervasive features of experience. In this character it is not so far different, after all, from a large range of important scientific thinking—like the theory of evolution, with its immense explanatory power that resists, more than Logical Positivist theorizers supposed, the decisive verifications, specific predictions, or the "crucial experiments" that turn out, often even in science, to be elusive.[7]

We may fairly conclude, then, that thinking by the "method of difference" is not necessarily the only responsible way of dealing theoretically with issues of great scope. But there remains a serious problem for ways of thinking that rest their explanatory power on larger and larger coherences and commend themselves for belief by pointing to the plausible patterns they can make out of all the facts. The problem is the acute possibility of generating several mutually incompatible but equally coherent and adequate patterns, each suggested by different models extended over the whole domain of reality, between which it seems theoretically arbitrary to choose. It is the old story of the Reason of Plato lacking a decisive *principle of choice* (4.1.e).

b. Values in Metaphysical Thinking

At this point there may be no hope for a purely theoretical solution. All the theoretical criteria are, we assume, equally satisfied in the imprecise way that criteria of such generality, dealing with domains of such complexity, can be on balance satisfied. Therefore there is nothing left on the theoretical level by which to make our choice. Theoretically, even if there were only one pattern that could satisfy our requirements at any given time, it would remain arbitrary why that particular pattern rather than some other should be the pattern of our actual universe. It could

---

[7]Thomas S. Kuhn, *The Structure of Scientific Revolutions* (Chicago, Ill.: University of Chicago Press, second edition, 1970).

easily have been some other way. As Whitehead admits, "In a sense, all explanation must end in an ultimate arbitrariness."[8]

What would allow the mental quest for comprehensive understanding satisfaction, if it is ever to find satisfaction, would be the *value* judgment that the pattern suggested by a metaphysical model of all reality is not only theoretically coherent and adequate but also right. Without this sense of rightness, the ultimate arbitrariness of explanation allows no peace. With it, a metaphysical scheme might offer a dwelling place for the mind and a coherent habitation for a unified life.

In this way, then, profound value judgments become involved with metaphysical thinking; that is, metaphysics is influenced by religious value-commitments no less than vice versa (7.5). Whatever part of the whole is taken as the clue to fundamental reality, whether it is the model of personal existence (e.g. theism), or mechanical processes (e.g. materialism), or life itself (e.g. organicism), it is implicitly offered as *normative* for reality. If one's values, like Einstein's, lead one to be skeptical of the ultimate worthiness of personal traits like love, knowledge, or the like, a theistic model as the basis for metaphysical thinking will not ring true; perhaps a more severe set of traits like precision, reliability, impassiveness, will lead one to prefer a mechanistic world model instead. In such ways, ultimate (religious) value-predilections may influence one's choice of metaphysical models. If, on the other hand, careful thinking by concepts articulating, say, an organismic model should successfully lead one to a particularly coherent and adequate theory of reality, such thinking may gradually enhance appreciation of, e.g., ecological values that had previously remained in the background. In these ways metaphysical models may reciprocally influence religious perceptions and sensitivities.

We can fairly conclude, then, that value judgments on this level are far from expressions of feeling alone. Our values do dispose us to feel, strongly at times, as circumstances arise. If we never felt emotions of approval or disapproval toward what we say we value, our sincerity might well be questioned. Likewise, our values incline us to behave, dramatically at times, as circumstances warrant. If we never acted to support what we say we value or to oppose what we claim to disvalue, it would be hard to hold that our values were genuine. Sometimes, in fact, we need to note what we actually do—how we spend our time, with whom, doing what, and how we spend our money—in order to clarify even for ourselves what our real values are. But normally we understand our values reasonably clearly through our beliefs about what is valuable. These beliefs are not logically identical to our feeling states, as the emotive theory of value holds. Two main differences prove their independence:

(1) My value-feelings and my value-beliefs *do not always vary together*, as they would have to if they were the very same thing. Sometimes I may feel strongly about something but do not have any particularly strong beliefs about the values

[8]Alfred North Whitehead, *Science and the Modern World* (The Macmillan Company, 1925; paperback edition, New York: The Free Press, 1967), p. 92.

involved; sometimes I may hardly feel anything, in a detached (or sleepy) mood, but be completely definite about my value beliefs; sometimes value-beliefs may painfully conflict with value-feelings, as Melville showed in *Billy Budd*, which situation poses a particularly poignant problem for value-actions. It is much too simple to equate value judgments with feeling-states or ejaculations of emotion.

(2) What makes value-beliefs most importantly different from value-feelings, even when they do happen to vary together, is the presence in the former of a universal element that must be missing from any simple feeling-state that, as a particular feeling, has no reference beyond its own time and place. If generosity, for example, is judged to be a valuable trait, it is not just this or that instance of generosity that is being referred to. The judgment refers to all cases that can properly be so described. Likewise if honesty is judged to be morally right, the judgment implies a universal: that I (or everyone relevantly like me) should act with honesty even when I (or they) do not feel like it. This universal element in value-beliefs makes it logically possible to argue over value-judgments. Such arguments do not have to be acrimonious. They do not even need to be with someone else. I can argue with myself over whether I should value, say, honesty over generosity, or the other way around. I can review with others the quality of our joint moral thinking—how completely we have formulated the issues, how consistently we have drawn our consequences, and the like. It is false to suppose that ethical arguments, even when there are important disagreements, must reduce to shouting matches.

Value-judgments, however, despite their openness to reasonable discussion of this sort, do not lend themselves to coercive settlement. We may focus and clarify the questions, bring matters back to the basic and nearly universal human goods (6.1.a), pose alternatives, and draw attention to full consequences—but in the end the personal freedom to make a contrary judgment in priorities, to see the issues differently, remains. In many ways that is a precious freedom. But in that freedom to differ is also lodged our responsibility to differ (or agree) thoughtfully.

Summary

Metaphysical thinking is a blend of systematic, comprehensive *theoretical modeling of all reality*, on the one hand, with *basic valuing of some aspects of reality*, expressed through the models we choose, on the other. Sometimes the valuing is more explicit, sometimes more implicit. Sometimes value-judgments lead, sometimes they follow. As in all immensely general and complex theoretical issues, the theoretical criteria of consistency, coherence, and adequacy are appropriate but hard to apply in any clear or decisive way; as in all value-discussions, there is room for thoughtful assessment but there are no coercive proofs. People who need firm and early closure on issues, and those who have little tolerance for ambiguity, should probably stay away from metaphysical thinking. But for those who value rule-guided thinking in the broadest terms, sorting out the stronger alternatives even without the promise of final answers, finding a finite but suitable dwelling place for thought and life, the metaphysical quest remains an endless attraction.

There is no room in this book to develop or even discuss full-blown, comprehensive metaphysical theories of reality as a whole. Lesser metaphysical issues, however, dealing with selected aspects of the "really real," tie into the larger syntheses as examples, carrying crucial implications down from the whole to the part (1.3). Since we are especially interested in sampling the ways that images drawn from technology may be adopted as metaphysical models for theorizing about reality in general, we shall content ourselves here with a few glimpses of the way in which these may have important bearing on our beliefs and attitudes toward human nature.

8.3. TECHNOLOGY AND MODELS OF HUMAN NATURE

People tend to be interested in themselves, in what they really are, in what "makes them tick." One of the most fascinating implications of any general metaphysical theory of reality, therefore, will be what it says about human nature. That will touch the sensitive point where our most intimate self-image and our most comprehensive theories about the world as a whole come together.

The technology of clocks, and machines in general, provided an immensely influential model for modern thinking about reality in general. Long before Newton's scientific depiction of the astronomical and physical order as a great machine, the regular motions of the moon and planets had been modeled in clockworks. Great European cathedrals, like Strasbourg, combined the display of religious and astronomical regularities in their awe-inspiring clocks, some of which still function for the admiration of tourists today. The image of the clock could capture the metaphysical imagination. Its parts were exquisitely adjusted to one another, without redundant or conflicting elements, embodying cooperation, regularity, reliability, numerability, and intelligibility. One could see, with enough patience, what "made them tick." With Galileo and the stress on the language of mathematics as the key to "reading the book of nature,"[9] with Descartes and the integration of the geometry of space with the algebra of thought,[10] and with Newton's triumphant equations, the model of the clock-machine could be articulated in detail by the powerful mechanistic theory of the world.

This "mechanical world picture"[11] has obvious significance for human nature. If everything, at bottom, runs like clockwork, then it must follow that we, too, are complex and wonderful machines. Consistent with William Harvey's (1578-1657) contemporary discovery that the heart is a pump circulating blood through our bodies in a fluid-mechanical system of valves and pipes, Thomas Hobbes (1588-

---

[9]Galilei Galileo, *Il Saggiatore* (1623), cited from E.J. Dijksterhuis, *The Mechanization of the World Picture* (London: Oxford University Press, 1961), p. 362.

[10]*Descartes Selections*, ed. Ralph M. Eaton (New York: Scribner's, 1927).

[11]Dijksterhuis, *The Mechanization of the World Picture*, trans. C. Dikshoorn (London: Oxford University Press, 1961).

1679) drew the full implications of the mechanical world model even before Newton (1642-1727). Everything about us, Hobbes theorized, would turn out to be matter following the laws of motion, if only we could see the fine-grained particles of which we are made. Even our minds are made of material arranged, like complex clockwork, in a mechanical system. Our perceptions, ideas, reasonings, are "motions" produced by pressures coming in from outside our organism, which in turn produce outward-directed pressures when we "decide" to react to the world. Our decisions, like our perceptions, must be purely mechanical processes, "for motion produceth nothing but motion."[12] Hobbes has won many followers in the modern world he helped to shape, although this model of human nature omits qualities—freedom, the dignity of being something more than "mere" matter—that many people would readily value in themselves and others. We still hear echoes of this metaphysical theory of human nature in ordinary speech, when we use mechanical images—e.g., "That makes him tick," "I am really wound up today," "You had better get in gear," "She really has gotten up a head of steam," etc.—to describe human character and processes.

A different metaphysical conclusion within the same general framework was drawn by René Descartes (1596-1650), who agreed with his contemporary, Hobbes, that the human body is purely mechanical, but exempted the human mind from the material world picture altogether. Major values were preserved by Descartes's decision, especially human autonomy to decide issues free from mechanical forcing, the possibility of human immortality, the special dignity of rationality, and the chance to represent an "image" of God's spiritual nature. These values were paid for, however, by the resulting theoretical incoherence that threatens such dualistic perspectives: the lack of any conceivable connections between the material universe, on the one hand, running in its machinelike fashion only by pushes and pulls, and mental phenomena, on the other hand, operating only by persuasion and decision. How could the mind affect the body, to make it act as we want, if the mind is not itself a mechanical pusher or puller? How can the body's sense organs affect the mind, if the mind is nowhere "geared in" to the mechanisms of the body? These famous metaphysical puzzles, among the most persistent in (and definitive of) modern philosophy, need not be pursued here; but it is clear that they are in large part the legacy of technological metaphysical models. Value preferences as well as theoretical arguments, as we have seen, will influence the debate over which models to choose and how far to push them; and our dominant theories of what things are like will in turn influence the ways we value—and treat—each other, and how we relate to the world of nature.

In more recent times the metaphysical debate over human nature has been much influenced by new images. Just as physics no longer limits its concepts and theories to the traits of corpuscular "matter in motion," but includes energies and processes of additional sorts—even analyzing matter itself into pulsations of energy—so the dominant technology for current imagination is not the mechanical clock but the

[12]Thomas Hobbes, *The Leviathan* (1651, edited by Molesworth, 1839–45), Part I, chap. 1, p. 2.

electrical computer. This, too, shows itself in current ways of speaking: e.g. "He was programmed to do it," "That just doesn't compute," "They want to interface their ideas with ours," "Let me process this a bit more."

It is too soon to draw many conclusions about the probable results of the leaping imagination of our time. Some enthusiasts can leap far indeed. John Haugeland describes the conceptual foundations of the new quest for Artificial Intelligence bluntly:

The fundamental goal of this research is not merely to mimic intelligence or produce some clever fake. Not at all. "AI" wants only the genuine article: *machines with minds*, in the full and literal sense. This is not science fiction, but real science, based on a theoretical conception as deep as it is daring: namely, we are, at root, *computers ourselves.* [13]

Again values as well as theoretical criteria will powerfully influence the debate. Computers do some things wonderfully well. The more that human beings find themselves convinced that they *are* indeed computers, the more they may be a tendency to stress those traits—inference-drawing, calculation, dispassionate rationality—as the appropriate ideals in which to define our humanity. This, if it happens, will surely have immense consequences for ethics and religion, for society, art, literature, and all that human beings touch or do.

On the other hand, it may turn out as computers become more and more intelligent in the ways they can be said to be intelligent, people will define themselves *in contrast* to what computers can do better than they. If my cheap pocket calculator does better at multiplication than I do, then I may not wish to define my essential self as second-fiddle to a silicon chip. Some may once have been tempted to define "rationality" as the capacity to calculate; but, others may object, it has become evident from technologies never previously imagined that number-crunching and other symbolic manipulations can no longer be taken as the unique mark of the human. Instead, it may be urged, the emotional side of the human—the side that can rhapsodize over a sonnet, reverberate to a symphony, or soar in prayer—should be identified with human nature.

What are we like? What is the universe like? These are never finally answered, but never-ceasing, questions. The character of our technologies and our attitudes toward those technologies provide us with hints and clues, with foci for preferences and patterns for thinking. What we come to think, however, as a society as well as individuals, *will have consequences.* This is another reason why, even though we cannot expect to force unanimity on such matters, it is vital that we think on these topics with the utmost care.

[13] John Haugeland, *Artificial Intelligence: The Very Idea* (Cambridge, Mass.: The MIT Press, 1985), p. 2.

8.4. FREE WILL AND TECHNOLOGICAL DETERMINISM

Another concern rooted in beliefs about human nature and technology, one that surfaces not only among philosophers and theologians (7.3 and 7.4) but also in popular culture—in film and novel and cartoon humor—is the question whether technology is really under human control or whether the machines have taken over. "Who is in charge here?" Does technology have such a powerful inner dynamic that we are merely fooling ourselves when we insist that "mere tools" could never rule their makers? Is human autonomy an illusion? Are the futures of modern society, and within society our individual lives, determined willy-nilly by technological imperatives that will override all human attempts to change course?

This issue, like all metaphysical questions, is part of a comprehensive theory of reality. At one step higher, the issue of technological determinism versus human autonomy is a special case of the more general metaphysical question whether human beings can be thought to have the sort of freedom of action that can genuinely initiate events, not merely react to causal pressures as the Hobbesian model requires. At a yet higher level of comprehensiveness, this intermediate metaphysical question is inevitably a special case of whether there exists at all, in the whole of reality, the sort of agency that can begin or purposefully redirect causal chains, not merely serve as one more link inside them. If the answer at the most general level is negative, then human autonomy would be ruled out *a fortiori* and determinism of some sort would be inevitable, though it would require a separate argument to show that technology, rather than something else, is the principal or sole determiner.

If we consider the issue at its most general level, none of the characteristic metaphysical models of modern technological thinking offer much basis for theory-articulation that would include anywhere in the universe a rich conception of causal autonomy. There are senses in which the word "freedom" can be used, of course, even of clockwork. The hands of a clock can move "freely" as long as nothing is obstructing them. The gears and springs inside are thereby being "freely" expressed. Despite such cosmetic language—called "soft determinism"—it is clear that a universal theory of reality modeled on mechanisms of this sort will have no theoretical place for the sort of event-initiating causal autonomy that is of interest here. In a clockwork universe, forces are passed along but never initiated. The last thing we expect—or want—from a clock is creativity.

What, though, of computers? Could a newer metaphysical model based on the capacities of intelligent machines provide a foothold for creativity in the sense we are looking for? Even now computers are capable of many surprising things. Poetry and painting, music composition, and superior chess moves are generated from what Haugeland calls "a bunch of general information and principles, not unlike what teachers instill in their pupils."[14] As future research develops, we may expect much

---

[14]*Ibid.*, p. 12.

more that is both unpredictable and brilliantly apt. This is a far cry from clockwork. True, but even this degree of creativity—and it is a genuine sort of creativity that human beings would often be grateful for—is wholly analyzable into the fully determined processing of previously available materials. It is not causal autonomy or the freedom to create (initiate or oppose) lines of influence. In metaphysical theory modeled on computer technology, the whole meaning of the terms in our discussion—"freedom," "agency," "creativity"—is exhausted by the sort of surprising intelligence of which computers are and will be increasingly capable. The kind of freedom that provides a living alternative to total determination by past causal circumstances is not part of that meaning. The logical geography of the universe permitted by such a model contains no place for such a concept. On that model it is unthinkable. Therefore, if our initial quest should be for exactly that sort of "unthinkable" human autonomy, and if we were given only modern technological models of reality to work with, then determinism would triumph by default.

In such a deterministic universe, is it technology itself that is now pulling the levers of historical change? Are our artifacts no longer merely tools but, rather, to be seen as our masters, running out of human control?

The primary evidence for this metaphysical vision would seem to be the extent to which vital human interests are regularly sacrificed under the wheels of the advancing technological Juggernaut. *Item:* Human beings need clean air and pure water for the basic goods of survival and health (6.1.a), but technological development fouls the air and pollutes the water. Attempts to control environmental damage are made with ever more technology, rather than simple human acts renouncing the damaging technologies, but each "technical fix" leads to another round of problems calling for ever more technological fixes. *Item:* Justice requires the fair distribution of harms and benefits (6.1.b), but mighty technologies of centralization, public and private, grow anyway, forcing nuclear or chemical plants—or waste storage dumps—on protesting local inhabitants in order to spread marginal benefits to distant, anonymous multitudes. *Item:* The goods of personal dignity and participatory freedom call for privacy and shared control, but the booming computerized society increasingly invades everywhere and makes argument with bureaucracies increasingly futile. *Item:* Survival, health, material security, society itself call for the dismantling of nuclear, chemical, and biological technologies of war that can and (in enough time) will destroy everything that is of human value, but the war machines continue to devour our treasure and dominate our policies.

One does not need to take the matter as far as Ellul (7.3) to see that there is a case to be made for technological determinism. From the standpoint of modern technological metaphysical models there is nothing to be said against determinism in principle, and the momentum of technological society seems unresponsive to human protest and even basic human needs. Is this gloomy word the last that philosophy of technology can say on the subject?

8.5. TECHNO-SCIENTIFIC THINKING AND ALTERNATIVE METAPHYSICS

From the viewpoint of this book there must be more to be said, since technologies themselves were defined as practical implementations of intelligence—*human* intelligence. If this understanding is at all near the mark, then certain logical consequences follow. First, it follows that technologies, no matter how threatening, *cannot ever be alien* to humanity. We may not like what we see when we look in our bathroom mirror, but the mirror both reflects and is human reality. Our human intelligence and human values—for better or for worse—are incarnate in our technologies. If some practical implementations of intelligence are brilliant and beneficent, to that degree must our species be so as well. If some are stupid or evil (or, worse, clever and evil), humanity is stupid and clever and evil to that extent, too. Second, it follows that our technologies, as practical methodologies, will tend to perpetuate themselves (3.3) once are established. Call it the "conservation of mental energy," the "conservatism of methodological intelligence," "social inertia"—by any name we choose, we should expect to find that major methods are difficult to change, that they become entrenched, develop a momentum and a constituency, and that their defenders refuse "to speculate freely on the limitations of traditional methods."[15] Perhaps our often painful experience of human mentality's own institutionalized inertial resistance—what Whitehead regretted as "obscurantism"— is the basis for our sometimes despairing sense that our technologies are beyond human control.

What makes the case even more complex is that our most problematic technologies rise out of the union of practical with theoretical reason. Modern technologies are the practical implementations of techno-scientific thinking, which itself constitutes a methodology, a logic, or a mode of consciousness with its own inertial resistance to change. As we noted earlier (4.4), modern techno-scientific thinking, despite its vast success, has certain flaws. These we summarized as *incoherence*, or the failure to achieve synthesis for understanding, and *inadequacy*, or the failure to include subtle data in the powerful but ideally simplified concepts and models it uses.

Exactly these flaws in techno-scientific thinking are the ones embodied in the technologies that threaten other human values today. First, by the incoherence of modern techno-scientific thought is meant its drive to specialization and analysis, splitting up problems for piece-meal attack rather than weaving solutions together into differentiated wholes. Technologies that implement this sort of thinking will inevitably have "side effects" that bring about chain reactions of further unwanted and unanticipated "side effects." Thinking synoptically in terms of wholes, however, there are in nature no "side effects," there are only *effects*, some of which are desired and some not. But modern techno-science threatens the environment even

---

[15] Alfred North Whitehead, *The Function of Reason* (Boston, Mass.: Beacon Press, 1929), p. 43.

with its successes because it takes sight on its problems with what Barry Commoner calls "tubular vision."[16]

Second, by the inadequacy of techno-scientific thinking is meant its tendency to ignore or even deny non-quantifiable aspects of experience and to force the reduction of subtle intuitions, like personal freedom (in the sense of self-determining agency[17]), into the inhospitable framework of metaphysical models that allow them no room. From such methods of thought come modern technologies insensitive to the vital non-quantitative requirements of human life, like beauty and justice. And from such forms of consciousness come implements, both mechanical and electronic, with capacities to overwhelm individual freedoms and dignities in the interests of impersonal efficiencies.

The biggest philosophical question of our age, then, is whether there might be some alternative to modern techno-scientific thinking. The critical gadfly of the Reason of Plato is able to sting at the inadequacies of present dominant methodologies (4.3 and 4.4), for reasons we have given, and many more,[18] but can anything positive be suggested?

One suggestion well worth exploring is that we try to draw our metaphysical models from ecology, perhaps in the form of an ecosystem, rather than from the physics-based technological models of the modern world. In several ways the field of ecology bids fair to become the leading "post-modern" science, since it does not share the basic metaphysical and epistemological assumptions, characteristic from the time of Galileo, of the modern thought-world.[19]

First, ecology cannot do its scientific work without an emphasis on *synthesis* in thinking. This does not mean, of course, that ecology abandons analysis. Not at all. Ecology is not *pre*-modern in its epistemology but *post*-modern, including the powers of modern scientific analysis in its repertoire but demanding that once the parts are clarified, they be seen again as critically differentiated wholes. Ecology is free to be *coherent* in its thinking, recognizing everything as connected to everything else.[20]

Second, ecology cannot do its scientific work without acknowledging the rich spontaneity, the purposes and values, of the living world which is its subject matter. It is not required to explain away or deny such phenomena as do not appear in a ball rolling down an inclined plane. It is free to be *adequate* in framing its concepts and its models.

Third, ecology cannot do its scientific work without self-reference. Ecologists

---

[16]Barry Commoner, *The Closing Circle: Nature, Man & Technology* (New York: Knopf, 1971; Bantam Books edition, 1972), p. 182.

[17]Frederick Ferré, "Self Determinism," *American Philosophical Quarterly 10*, no. 3 (1972), pp. 286–304.

[18]Frederick Ferré, *Shaping the Future: Resources for the Post-Modern World* (New York: Harper & Row, 1976), especially pp. 1–67.

[19]Frederick Ferré, "Religious World Modelling and Postmodern Science," *Journal of Religion 62*, no. 3 (July, 1982), pp. 261–71.

[20]Barry Commoner, *The Closing Circle*, p. 29.

are a small but significant part of the natural whole that ecologists study. The human and the natural order are not viewed as separate and alien, in ways that may encourage the attempt to dominate and control as though humans were "outside" nature. Instead, the human and the natural are inseparably intertwined—so much so that they neither exist, nor can any longer be fully understood, apart from each other.

In such ways, it may be speculated, a significantly different mode of thinking, a full post-modern metaphysical scheme allied to post-modern forms of science, may be a candidate to replace modern techno-scientific thinking with a new form of intelligence. This form would be more coherent, more adequate, and more oriented to life. This is high-flying speculation. Whitehead and his followers have done much to spell out such a "philosophy of organism," as Whitehead called it; but, although serious efforts are being made toward articulating such a scheme in terms of contemporary knowledge,[21] little embodiment of the scheme has made its way into contemporary institutions—the "supreme authority" for such speculative flights (4.4). High-flying is what the Reason of Plato does best. This calls for no apology. But perhaps we can come at least part-way back to earth with some final speculations about the implications of such a post-modern metaphysics for the future of technology.

8.6. POST-MODERN THINKING AND THE FUTURE OF TECHNOLOGY

Three points need to be kept in mind as we approach our final speculations.

First, no matter how the future of civilization may differ from the present or from any past period, there will be technologies if there is civilization at all. It may not remain "*our* civilization" if there are too many discontinuities with the modern period, in whose later phases (perhaps) we are now living. This should not be shocking. Civilizations rise and flourish; they also fade or convulse. Ours has no guarantee of immortality. On the contrary, its mortality is evident. Our proudest achievements, our typical technologies, seem likely to destroy us unless we change their character—hence our civilization's own deepest character—in swift and drastic ways.

Second, it may be that science need not be modern science to be real science. Just as there were pre-modern sciences, perhaps there could be post-modern sciences as well. One example given was ecology as a potential post-modern science: critical, penetrating, empirical, and yet not wholly comfortable with the approaches of analysis, reduction, and alienation that have been the marks of modern science. Other examples might be drawn from other recent developments in the sciences.[22] What should be clear is that the outright identification of science itself with the

---

[21]David R. Griffin, *Physics and the Ultimate Significance of Time* (Albany, N.Y.: State University of New York Press, 1986).

[22]See Harold K. Schilling, *The New Consciousness in Science and Religion* (Philadelphia, Pa.: A Pilgrim Press Book, 1973); Ian G. Barbour, *Myths, Models and Paradigms* (N.Y.: Harper & Row, 1975).

metaphysical and methodological models and habits with which it has lived closely for the past four centuries should not be granted without challenge.

Third, technology need not be modern technology in the sense of being more of the same, just bigger and faster. As there were pre-modern technologies, perhaps there may be post-modern, radically different technologies. It is always tempting to extrapolate trends in straight lines into the unknown future, but historical change is seldom so simple. If there should be an "organismic" post-modern civilization with its own characteristic ways of thinking, its own metaphysical models, and its own forms of science, then future technologies will be post-modern too.

If we try to imagine, in an appropriately general way, what an organismic metaphysical and institutional framework would mean for post-modern technologies, the first thing we notice is that such technologies will aim at *optimization* rather than *maximization*. Healthy organisms and populations need homeostatic restraints. Bigger is not always better for an organism, and more is not always a healthy goal. These principles will make a vast difference to the character of artifacts and the system of economics in such a post-modern world. Systems of production will not automatically aim for maximum efficiency or profit. Stability, durability, sustainability, and satisfaction will be dominant considerations.

Second, organismic thinking will lead to more technologies of *cultivation* and fewer of *manipulation*. Instead of adopting an externalist attitude toward nature, post-modern institutions like ecological agriculture will attend to internal biological rhythms and ecosystem restraints in food production. Pouring on petrochemicals will give way to methods of nurture. Regenerative farming, with full attention to the needs of the land and its biota, would be the natural alternative to the energy-intensive, resource-depleting, variety-threatening, pollution-producing agribusiness of the modern era.

Third, post-modern thinking of the sort we are considering will embody technologies of *differentiation* rather than *centralization*. Healthy organisms are mutually differentiated, internally related systems of information and energy. The parts participate in maintaining the whole and the whole benefits the parts. The personal computer, expressing the individuality of its user, but linked with others in larger and larger networks of interactive communicators, would be a good symbol for post-modern technology. So would be a cooperative interchange of electrical power among large numbers of small solar and wind generators, replacing the centralized power plant and the monopolizing electrical grid.

In the end, of course, our speculations may only be so much utopian romance. The modern world may collapse with no such benign successor, or it may be surprisingly tough and may manage to survive into the indefinite future. Organismic models may, despite efforts, fail to articulate sufficiently coherent and adequate theories of reality as to warrant respect. Even ecology itself may eventually become a domain for reducers and analyzers, settling for parts rather than wholes and thus gaining—along with quicker experimental results and increased grant money—"modern" respectability in the eyes of the rest of the scientific establishment.

This is not a book of prophecy, just an essay in the philosophy of technology. But the best way of seeing one's own country, they say, is to return to it from abroad. Then features that one might never have noticed—especially the pervasive features—come clear with the power of culture shock. Modern technology is where we live, in a technosphere that is immensely hard to avoid. This book has tried to provide a trip into the foreign lands of conflicting values and many unfamiliar tongues, so that now, at the end, a return home to modern technological civilization will permit all who were on this voyage to see their starting place—its blemishes and its delights—as never before.

Glossary of Selected Terms

adequacy The measure of a theory's or a concept's capacity to reflect data accurately and fully.

aesthetics The branch of philosophy that studies beauty in general and the experience of art in particular; a sub-field of axiology.

AI The field of research into artificial intelligence, especially the development of computers capable of some symbol-using behaviors indistinguishable from the human.

analysis Reasoning by dividing the problem into its smallest component parts, especially the philosophical examination of issues by minute examination of the language in which the problem is expressed.

anarchical Unruled, unruly; especially (for Whitehead) appetitions not under the control of reason.

anthropocentrism Centering around the human, especially valuing humanity as centrally or exclusively important.

anti-entropic Tending toward increasing order or available energy; see entropic.

appetition Desire for or seeking after something.

articulate To put into speech or, especially, to give formal structure to something, particularly to a model by a theory.

artifact Something made at least in part by "art" or intelligence.

axiology Philosophical study of values in general.

beneficence Making or supporting goods.

benevolence   Willing or wishing well.

calculus   System of symbols, with rules for connection, separation, and transformation; see theory.

category   Major concept in a theory.

charlatanism   Pretension to knowledge, especially (for Whitehead) the fate of speculative reason undisciplined by practice.

coherence   Measure of a theory's capacity to provide internal connections among its categories and lesser concepts.

comprehensiveness   Measure of a theory's scope or inclusiveness.

concept   A complex of abstract characteristics.

concreteness   Measure of a theory's or concept's capacity to approximate the full actuality of its subject matter; the full actuality of something itself.

consistency   Measure of a theory's freedom from internal contradictions.

contradiction   Statement that denies itself.

data   Contents given for a theory to deal with.

desublimation   Narrowing the gap between the desired and the permitted, thus (for Marcuse) undermining criticism of modern society.

determinism   The view that all events are exactly required by previous causal conditions.

emotive   Involving an expression of feeling; particularly the analysis of values, ethical or aesthetic, as nothing but ejaculations of feeling.

empirical   Based on experience.

entropic   Tending to the decay of order or loss of available energy.

epistemology   Philosophical study of knowledge in general.

ethics   The branch of philosophy that studies morality in general, particularly goods and the right; field of axiology.

ethnocentrism   Centering around one's own culture, especially considering the values of one's own group to be centrally or exclusively important.

extrinsic   External to something; especially values that are justified by something other than themselves; see intrinsic.

falsifiable   Open to empirical challenge.

gadfly   Disturber or critic of complacency.

goods   The aims of appetitions.

harms   Destruction or withholding of goods.

homeostatic   Characterizing biological processes that maintain equilibrium.

hubris   Excessive pride.

implemented   Made actual, usually with artifacts.

intelligence   Capacity for self-disciplined mental activity.

intrinsic   Internal to something; especially values that are justified in terms of themselves alone, like happiness; see extrinsic.

intuition   Direct awareness of something.

justice   Principled avoidance of discrimination in distribution of goods and harms unless warranted by morally relevant differences in circumstances; see merit.

logic   Self-discipline of theoretical intelligence.

machine   An artifact made of separate interconnected parts working together for some common effect.

materialism   The philosophical view that reality is best interpreted according to the model of material objects; also, valuing material things as centrally or exclusively important.

mechanism   The philosophical view that reality is best interpreted according to the model of a machine.

mental   The capacity to take account of what is not concretely present.

merit   Circumstances offered as a morally relevant justification for favorable discrimination in keeping with justice.

metaphysics   Philosophical study of reality in general.

methodology   Philosophical study of method in general, including basic definition and establishment of categories.

model   Any representation; especially an interpretation of some further subject matter for a theory.

non-maleficence   Avoiding doing harms.

obscurantism   Refusal to inquire critically, especially (for Whitehead) about dominant methods in use.

ontological   Dealing with being.

organicism   The philosophical view that reality is best interpreted according to the model of living organisms.

phenomenology   Philosophical description of general structures of experience.

reason   Mentality under self-discipline; see intelligence.

Reason of Plato   Theoretical intelligence, especially (for Whitehead) speculative reason.

Reason of Ulysses   Practical intelligence, especially (for Whitehead) practical reason.

reduction   Statement of one subject matter entirely in terms of another that is considered on a lower order of organization.

speciesism   Giving unwarranted preference to one's own species; see anthropocentrism.

standing-reserve   Reduction of all things to manipulable utilities; especially (for Heidegger) the effect of the essence of technology.

synthesis   Reasoning by connecting concepts into coherent wholes.

techne̅   Greek term for arts or skills in general.

technique   Rationalized method, especially (for Ellul) the autonomous quest for perfect efficiency characterizing modern technological society.

technocracy   Rule by technicians, engineers, and scientific experts.

technology   Practical implemented intelligence.

theory   A formal calculus of symbols in systematic relations; if explanatory, these symbols require interpretation as concepts for a subject matter, often through a model which is thereby articulated.

verifiable   Open to empirical confirmation or support.

Suggestions for Further Reading

The list below includes selected works, most of which provide extensive further references to the literature in the field.

GENERAL ANTHOLOGIES AND SERIES

Conley, Verena Andermat, ed., *Rethinking Technologies*. Minneapolis: University of Minnesota Press, 1993.

Kranzberg, Melvin, and W. H. Davenport, eds., *Technology and Culture: An Anthology*. New York: Schocken, 1972.

Mitcham, Carl, and Robert Mackey, eds., *Philosophy and Technology: Readings in the Philosophic Problems of Technology*. New York: The Free Press, 1972.

Research in Philosophy and Technology. Greenwich, Conn.: JAI Press (an annual series since 1978).

Technology and Culture. Chicago: University of Chicago Press (series since 1959).

Teich, Albert H., ed., *Technology and the Future*, 4th ed. New York: St. Martin's, 1986.

SPECIALIZED ANTHOLOGIES

Baier, Kurt, and Nicholas Rescher, eds., *Values and the Future: The Impact of Technological Change on American Values*. New York: The Free Press, 1969.

Haigerty, Leo J., ed., *Pius XII and Technology*. Milwaukee, Wisc.: Bruce Publishing, 1962.

Johnson, Deborah G., and John W. Snapper, eds., *Ethical Issues in the Use of Computers*. Belmont, Calif.: Wadsworth Publishing, 1985.

Mitcham, Carl, and Jim Grote, eds., *Theology and Technology: Essays in Christian Analysis and Exegesis*. Lanham, Md.: University Press of America, 1984.

Truitt, Willis H., and T. W. Graham Solomons, eds., *Science, Technology, and Freedom*. Boston: Houghton Mifflin, 1974.

BOOKS BY INDIVIDUAL AUTHORS

Adas, Michael, *Machines as the Measure of Men: Science, Technology, and Ideologies of Western Dominance*. Ithaca: Cornell University Press, 1989.

Barbour, Ian G., *Ethics in an Age of Technology: The Gifford Lectures 1989–1991*, vol. 2. San Francisco: HarperCollins, 1993.

———, *Science and Secularity: The Ethics of Technology*. New York: Harper & Row, 1970.

———, *Technology, Environment, and Human Values*. New York: Praeger, 1980.

Barrett, William, *The Illusion of Technique: A Search for Meaning in a Technological Civilization*. Garden City, N.Y.: Anchor Press, 1978.

Barry, John A., *Technobabble*. Cambridge, Mass.: MIT Press, 1991.

Bunge, Mario Augusto, *Philosophy of Science and Technology*. Dordrecht, Holland/Boston, USA: Kluwer Academic, 1985.

Channell, David F., *The Vital Machine: A Study of Technology and Organic Life*. New York: Oxford University Press, 1991.

Cox, Harvey, *The Secular City: Secularization and Urbanization in Theological Perspective*. New York: Macmillan, 1965.

DeGregori, Thomas R., *A Theory of Technology: Continuity and Change in Human Development*. Ames: Iowa State University Press, 1985.

Drengson, Alan R., *Beyond Environmental Crisis: From Technocrat to Planetary Person*. New York: P. Lang, 1989.

Ellul, Jacques, *The Technological Society*, John Wilkinson, trans. New York: Knopf, 1964.

———, *The Technological System*, Joachim Neugroschel, trans. New York: Continuum Publishing, 1980.

Feenberg, Andrew, *Critical Theory of Technology*. New York: Oxford University Press, 1991.

Feibleman, James K., *Technology and Reality*. The Hague, Holland/Boston, USA: Kluwer, 1982.

Ferkiss, Victor C., *Nature, Technology, and Society: Cultural Roots of the Current Environmental Crisis*. Albany: State University of New York Press, 1993.

———, *Technological Man: The Myth and the Reality*. New York: New American Library, 1970.

———, *The Future of Technological Civilization*. New York: Continuum Publishing, 1980.

Ferré, Frederick, *Hellfire and Lightning Rods: Liberating Science, Technology, and Religion*. Maryknoll, N.Y.: Orbis Books, 1993.

———, *Shaping the Future: Resources for the Post-Modern World*. New York: Harper & Row, 1976.

Florman, Samuel C., *The Existential Pleasures of Engineering*. New York: St. Martin's Press, 1976.

Fuller, R. Buckminster, *No More Secondhand God and Other Writings*. Garden City, N.Y.: Doubleday, 1963.

Gendron, Bernard, *Technology and the Human Condition*. New York: St. Martin's, 1977.

Germain, Gilbert G., *A Discourse on Disenchantment: Reflections on Politics and Technology*. Albany: State University of New York Press, 1993.

Heidegger, Martin, *The Questions Concerning Technology and Other Essays*, William Lovitt, trans. New York: Harper & Row, 1977.

Hickman, Larry A., *John Dewey's Pragmatic Technology*. Bloomington: Indiana University Press, 1990.

Ihde, Don, *Existential Technics*. Albany: State University of New York Press, 1983.

———, *Instrumental Realism: The Interface Between Philosophy of Science and Philosophy of Technology*. Bloomington: Indiana University Press, 1991.

———, *Postphenomenology: Essays in the Postmodern Context*. Evanston, Ill: Northwestern University Press, 1993.

———, *Technics and Praxis: A Philosophy of Technology*. Boston Studies in the Philosophy of Science, vol. 24, Robert S. Cohen and Marx W. Wartofsky, eds. Dordrecht, Holland/ Boston, USA: D. Reidel Publishing, 1979.

———, *Technology and the Lifeworld: From Garden to Earth*. Bloomington: Indiana University Press, 1990.

Illich, Ivan, *Tools for Conviviality*. New York: Harper & Row, 1973.

Kohák, Erazim, *The Embers and the Stars: A Philosophical Inquiry into the Moral Sense of Nature*. Chicago: University of Chicago Press, 1984.

Levinson, Paul, *Mind at Large: Knowing in the Technological Age*. Greenwich, Conn.: JAI Press, 1988.

Marcuse, Herbert, *One-Dimensional Man: Studies in the Ideology of Advanced Industrial Society*. Boston: Beacon Press, 1964.

Marx, Leo, *The Machine in the Garden: Technology and the Pastoral Ideal in America*. New York: Oxford University Press, 1964.

Mesthene, Emmanuel G., *Technological Change: Its Impact on Man and Society*. Cambridge, Mass.: Harvard University Press, 1970.

Moore, D. J. Huntington, *A Metaphysics of the Computer: The Reality Machine and a New Science for the Holistic Age*. San Francisco: Mellen Research University Press, 1992.

Mumford, Lewis, *Technics and Civilization*, 2nd ed. New York: Harcourt, Brace & World, 1963.

Norman, Donald A., *Things That Make Us Smart: Defending Human Attributes in the Age of the Machine*. Reading, Mass.: Addison-Wesley Publishing, 1993.

Pacey, Arnold, *The Culture of Technology*. Cambridge, Mass.: MIT Press, 1983.

Rapp, Friedrich, *Analytical Philosophy of Technology*, Stanley R. Carpenter and Theodore Langenbruch, trans. Dordrecht-Holland/Boston-USA: D. Reidel Publishing, 1981.

Rescher, Nicholas, *Unpopular Essays on Technological Progress*. Pittsburgh: University of Pittsburgh Press, 1980.

Romanyshyn, Robert D., *Technology as Symptom and Dream*. London: Routledge, 1989.

Rosenbrock, H. H., *Machines with a Purpose*. New York: Oxford University Press, 1990.

Ross, Andrew, *Strange Weather: Culture, Science, and Technology in the Age of Limits*. London: Verso, 1991.

Roszak, Theodore, *The Making of a Counter Culture: Reflections on the Technocratic Society and its Youthful Opposition*. Garden City, N.Y.: Doubleday, 1969.

———, *Where the Wasteland Ends: Politics and Transcendence in Postindustrial Society*. Garden City, N.Y.: Doubleday, 1973.

Rothenberg, David, *Hand's End: Technology and the Limits of Nature*. Berkeley: University of California Press, 1993.

Schumacher, E. F., *Small is Beautiful: Economics as if People Mattered*. New York: Harper & Row, 1973.

Shrader-Fréchette, Kristin, *Risk Analysis and Scientific Method: Methodological and Ethical Problems with Evaluating Societal Hazards*. Dordrecht, Holland: D. Reidel, 1985.

Sikorski, Wade, *Modernity and Technology: Harnessing the Earth to the Slavery of Man*. Tuscaloosa: University of Alabama Press, 1993.

Tierney, Thomas F., *The Value of Convenience: A Genealogy of Technical Culture*. Albany: State University of New York Press, 1993.

White, Lynn, Jr., *Medieval Technology and Social Change*. New York: Oxford University Press, 1962.

Winner, Langdon, *Autonomous Technology: Technics-Out-Of-Control as a Theme in Political Thought*. Cambridge, Mass.: MIT Press, 1977.

————, *The Whale and the Reactor: A Search for Limits in an Age of High Technology*. Chicago: University of Chicago Press, 1986.

Zimmerman, Michael E., *Heidegger's Confrontation with Modernity: Technology, Politics, and Art*. Bloomington: Indiana University Press, 1990.

Index

abacus, 24
Abraham, 3
Adam, 102, 107–110, 113
adequacy, 6, 51, 52, 76, 78, 120, 122, 125,
 132. *See also* Glossary of Selected Terms
aesthetics, 3, 7, 8, 24, 27, 35, 45, 49, 67, 72,
 76, 86, 89, 132. *See also* Glossary of
 Selected Terms
agency, personal, 77, 88, 89, 90, 129
agriculture, 15, 28, 37, 66, 93, 95, 102, 109,
 134
alchemy, 27
Allan, George, x
American Indians, 114
Anderson, Wyatt, x
animals, 17–18, 31, 85
Anscombe, G.E.M., 21n
anthropocentrism, 85, 102, 103, 112. *See also*
 Glossary of Selected Terms
apparatus, 44–46
appearance, 117–120
Arabic numerals, 15
Arabic science, 44, 47
Aristarchus of Samos, 45
Aristotle, 2, 4, 11, 43, 44, 65
Armstrong, David M., 118n
artifact, 27, 37, 44, 51, 93. *See also* Glossary
 of Selected Terms

artificial intelligence, 27, 128. *See also*
 Glossary of Selected Terms
artificiality, 18–19, 27–28, 94, 120
arts
 applied. *See* crafts
 fine, 24, 65, 128
astronomy, 44
automation, 87–89
automobile, 86
aviation, 1, 19, 27, 34, 50, 58–60, 61, 65,
 67–68, 98
axiology, 3, 4, 7, 8, 11, 75, 116. *See also*
 Glossary of Selected Terms
Ayer, A.J., 119n

Baier, Kurt, 141
Barbour, Ian G., 82, 84, 85, 113, 133n, 141
Barnhart, C.L., 14n
Benedict, Saint, 113
Beneficence, Principle of, 79, 83, 86, 88, 92,
 93, 113, 115. *See also* Glossary of Selected
 Terms
Berkeley, George, 118
biological technologies, 20, 27, 28, 94–96
Brannigan, Vincent M., 90n
Brown, Lester, 94
Buddhism, 24, 114, 115